T0282456

SpringerBriefs in Applied Sciences and Technology

Forensic and Medical Bioinformatics

Series editors

Amit Kumar, Hyderabad, India
Allam Appa Rao, Hyderabad, India

More information about this series at http://www.springer.com/series/11910

Samira Hosseini · Fatimah Ibrahim

Novel Polymeric Biochips for Enhanced Detection of Infectious Diseases

 Springer

Samira Hosseini
Department of Biomedical Engineering,
  Faculty of Engineering
University of Malaya
Kuala Lumpur
Malaysia

and

Center for Innovation in Medical
  Engineering, Faculty of Engineering
University of Malaya
Kuala Lumpur
Malaysia

Fatimah Ibrahim
Department of Biomedical Engineering,
  Faculty of Engineering
University of Malaya
Kuala Lumpur
Malaysia

and

Center for Innovation in Medical
  Engineering, Faculty of Engineering
University of Malaya
Kuala Lumpur
Malaysia

ISSN 2191-530X   ISSN 2191-5318   (electronic)
SpringerBriefs in Applied Sciences and Technology
ISSN 2196-8845   ISSN 2196-8853   (electronic)
SpringerBriefs in Forensic and Medical Bioinformatics
ISBN 978-981-10-0106-2   ISBN 978-981-10-0107-9   (eBook)
DOI 10.1007/978-981-10-0107-9

Library of Congress Control Number: 2015955880

Springer Singapore Heidelberg New York Dordrecht London

Printed on acid-free paper

Springer Science+Business Media Singapore Pte Ltd. is part of Springer Science+Business Media
(www.springer.com)

Preface

In the area of biosensing applications, variety of sophisticated systems from the most advanced classes have been established and presented to the scientific community. Nonetheless, only minor percentages of such complex techniques have opened their ways from the laboratory benches to the actual clinical practices. Such a clear statistic has to encourage scholars and subsequently the industry to devote more efforts toward the construction of new generations of analytical platforms that can revolutionize current biodiagnostic tools and serve the humanity with the enhanced point of care. This book is dedicated to synthesis, processing, and fabrication of novel polymeric platforms that can be used as an additional into the current diagnostic systems for improved biorecognition. Our newly developed platforms can also be used as an adaptive technique for fabrication of a new-generation of analytical kits with higher efficiency. Developed copolymeric materials offer permanent existence of controlled functional groups, which can be effectively used for analyte–surface interaction. Although proposed methodology enables bioreceptor platforms to detect various types of viruses, our main target analyte, in particular, was chosen to be dengue virus. Dengue fever is one of the most threatening mosquito-borne viral infections mainly widespread in tropical and subtropical regions. The major emphasis of this book is to establish a relationship between property, chemistry, and morphology of the surface with biosensing ability of the developed platform. It also has thoroughly addressed communal questions in the field of analytical systems based on heterogeneous biorecognition.

Acknowledgments

This research is supported by University of Malaya High Impact Research Grant (UM.C/625/1/HIR/MOHE/05) awarded by Ministry of Higher Education, Malaysia and University of Malaya Research Grant (UMRG: RP009A-13AET).

Contents

Abbreviations

AFM	Atomic force microscopy
AIBN	Azobisisobutyronitrile
APM	Scanning probe microscopy
BSA	Bovine serum albumin
DAP	1,2-diaminopropane
DENV	Dengue virus
DF	Dengue fever
DHF	Dengue hemorrhagic fever
DSS	Dengue shock syndrome
EDC	1-ethyl-3-(3-dimethylaminopropyl) carbodiimide
ELISA	Enzyme-linked immunosorbent assay
ES	Electron spectroscopy
FN	False negative
FP	False positive
GA	Glutaraldehyde
HMDA	Hexamethylenediamine
LoB	Limit of blank
LoD	Limit of detection
MAA	Methacrylic acid
MMA	Methyl methacrylate
Na_2HPO_4	Disodium hydrogen phosphate
NaH_2PO_4	Monosodium phosphate
NHS	N-hydroxysuccinimide
NTDs	Neglected tropical diseases
PBS	Phosphate buffer saline
PEI	Polyethylenimine
Poly(MMA-co-MAA)	Poly methylmethacrylate-co-methacrylic acid
SEM	Scanning electron microscopy
SPR	Surface plasmon resonance
TB	Toluidine blue
TEGDMA	Triethyleneglycol dimethacrylate

THF	Tetrahydrofuran
TN	True negative
TP	True positives
WCA	Water-in-air contact angle
WHO	World Health Organization
XPS	X-ray photoelectron spectroscopy

Chapter 1
Current Optical Biosensors in Clinical Practice

Abstract This chapter presents fundamental aspects in fabrication and development of biosensors and covers a brief explanation about different signal transduction techniques. The major emphasis of the chapter is on the optical biosensors, in particular enzyme-linked immunosorbent assay (ELISA). Four major protocols including direct, indirect, sandwich and competitive ELISA, applied in clinical practices for running conventional analytical ELISA, are described and compared in a great detail. Furthermore, determination of important parameters such as sensitivity, specificity, accuracy and limit of detection (LoD) is provided, which are essential in careful evaluation of the assay. The chapter also reviews the drawbacks of the current conventional analytical platforms and possible functionalization techniques as a solution to the existing limitations. The chapter ends with the investigation of the functionalized surfaces with different concentrations of generated functional groups and comparison between performances of applied functionalized platforms.

Keywords Optical biosensors · Surface functional groups · Analytical assays · Sandwich ELISA

1.1 Principle of Biosensors

Biosensors, the concise form of "biological sensors", are analytical devices that can detect biological elements of interest. Biosensor devices rapidly expanded their applications into various fields such as food quality control, clinical diagnostics, environmental monitoring, and healthcare [1–3]. Desirable biosensors must possess some beneficial features such as:

- Sensitivity
- High specificity toward the targeted analyte
- Stability under typical storage condition
- Accuracy of the response

© The Author(s) 2016
S. Hosseini and F. Ibrahim, *Novel Polymeric Biochips for Enhanced Detection of Infectious Diseases*, SpringerBriefs in Forensic and Medical Bioinformatics, DOI 10.1007/978-981-10-0107-9_1

- Reproducibility and minimized intraday variability
- Non-toxicity
- Cost effectiveness
- Portability

The major principle in biosensor devices relies on the conversion of biological response into the electrical signal by the aim of transducer. Generated signal is further amplified, processed and displayed as the final result of biorecognition. Signal transduction can be performed by different types of analytical system like electrochemical biosensors, optical biosensors, cyclic voltammetry, electrochemical impedance spectroscopy, surface plasmon resonance (SPR), potentiometric signal transduction, gravimetric, and thermal transduction [4–7].

1.2   Enzyme-Linked Immunosorbent Assay (ELISA)

Enzyme-linked immunosorbent assay (ELISA) from the family of optical biosensors is one of the most versatile immunoassay techniques in routine clinical procedure. ELISA has rapidly found variety of applications in different areas such as food industry, toxicology, immunology and plant pathology. As a plate-based assay, ELISA is specifically used for the quantification of the biomolecules such as peptides, proteins, antibodies, viruses and hormones [8–11].

Namely direct ELISA, indirect ELISA, competitive ELISA and sandwich ELISA are different protocols for conducting this assay (Fig. 1.1).

Direct ELISA is considered to be the simplest type of assay among the others (Fig. 1.1a). In this protocol, antigen is initially adsorbed on the plastic surface of the analytical kit. As a following step, a blocking agent is added to the well plate in order to block non-specific binding sites. Afterwards, an enzyme-linked antibody is introduced to the assay, from which the enzyme will subsequently be released and detected when the substrate is added to the assay. Although direct ELISA avoids cross-reactivity due to the absence of labeled secondary antibody, it requires the labeling of the primary antibody, which is not suitable for all kinds of primary antibodies.

Indirect ELISA involves two steps of binding process at which primary antibody from one side and labeled secondary antibody from the other side bind to the antigen (Fig. 1.1b). While this method is very convenient and leads to the signal amplification, the risk of cross-reactivity is high due to the non-specific binding between conjugate secondary antibody and other two biomolecules.

In competitive ELISA, the antigen of interest from the blood sample competes with the purified antigen immobilized inside the well plate in binding to the antibody (Fig. 1.1c). In this particular method, decrease in the generated signal is an indication for the presence of the antigens inside the blood sample. The major advantage of this method over other protocols is that the impure samples can be

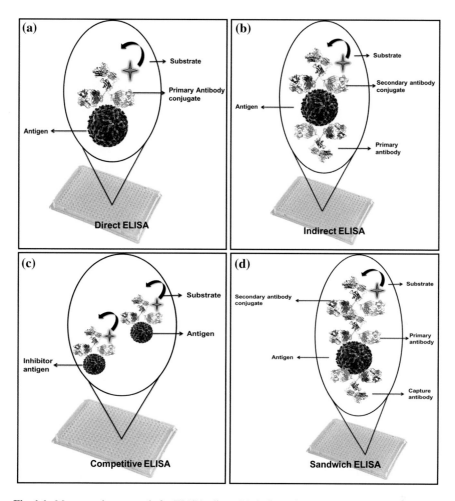

Fig. 1.1 Most popular protocols for ELISA: direct (**a**); indirect (**b**); competitive (**c**) and; sandwich (**d**) ELISA

used without compromising the reproducibility of the assay. Nonetheless, lower level of sensitivity and specificity makes this protocol less favorable.

Among different mentioned protocols for performing ELISA, sandwich ELISA has proven to be the most specific and reliable method for biomolecular recognition. In this strategy, which can be specifically used for antigen detection, the targeted antigen is sandwiched between primary and capture antibodies from both sides (Fig. 1.1d). These two antibodies are produced in different hosts' bodies thus they are incapable of binding one to another. Primary antibody however, can further react with the labeled secondary antibody, which leads to the emission of either fluorescence or color upon biorecognition. The detection signal intensity can be increased with the concentration of the antigen, proportionally. Antigens used in this protocol do not have to be purified prior to the assay but they must at least contain two antibody binding sites. Although the procedure is lengthy, it has proven to be sensitive and highly selective in comparison to the rest [12].

1.3 Sandwich ELISA

In this study, sandwich ELISA was selected as the protocol of choice as this method avoids uncertainties and minimizes the chance of non-specific binding. Detailed procedures of sandwich ELISA in a typical conventional assay are as follows.

1.3.1 Capture Antibody Immobilization

In the first step of sandwich ELISA, analytical kit is coated with capture antibody. The incubation is normally carried out for 2 h. The suitable temperature for such incubation is 37 °C, which is an equivalent temperature with the human body. Capture antibodies are typically diluted in coating buffer based on the specified guidelines given by manufacturers [13–15].

1.3.2 Washing Procedure

Washing procedure is one of the most important parts of the assay. A mixture of detergents such as 0.05 % Tween 20 with phosphate buffer saline (PBS) can be an appropriate buffer for washing step. The pH of such buffer is advices to be approximately 7.4. Washing procedure is normally performed at ambient temperature by using shaker (with shaking speed of 1000 rpm). This step must be conducted 3 times (each time at least 5 min) in order to make sure that the excess of residue has been thoroughly eliminated. The exact same procedure has to be repeated between each two steps of the ELISA [13–16].

1.3.3 Blocking Step

To achieve higher selectivity and to avoid non-specific binding, blocking procedure has to be performed as a necessary step of each assay. Among different compounds, bovine serum albumin (BSA) is known as one of the strong mediums for blocking the unreacted sites. Remained empty spaces between immobilized biomolecules, can be covered by blocking agent and expectedly the next introduced analyte can only bind to the previously immobilized biomolecule [13–16].

1.3.4   Antigen Attachment

Serial dilution is one of the known methods for preparation of the samples in clinical practices at which dilution factor is constant from one step to another. In this study, a ten-fold serial dilution was chosen as the standard protocol by which different concentrations of virus solution were prepared. Therefore, used virus solutions in the assay were either intentionally infected (positive) or non-infected (negative). The virus used in this study was propagated and isolated in our laboratories [14]. The incubation of the well plates with virus solution can typically be carried out for whichever 2 h at 37 °C or overnight at 4 °C [13–16].

1.3.5   Primary Antibody Attachment

In order to locally immobilize antigen between two antibodies, primary antibody should be introduced to the sequence as the next step. Similar to capture antibody immobilization, the incubation can be carried out for 2 h at 37 °C. This individual biomolecular entity, as described in Sect. 1.2, can further bind to the conjugate secondary antibody and subsequently release the detection signal [13–16].

1.3.6   Secondary Antibody Attachment and Reading the Signals

In the final step of ELISA, secondary and primary antibodies can bind one to another in the incubation, which is normally conducted for 30 min at 37 °C. In colorimetric ELISA (the protocol of choice in this study) a mixed substrate has to be added to the wells after final washing step. This incubation last for 10–15 min. Eventually stopping buffer should be added to the assay after which the detection signal can be recorded in the specific wavelength by using ELISA reader [13–16].

1.4   Evaluation of ELISA

It is highly important to fully assess the analytical performance of the newly developed clinical laboratory test in order to determine its capability and limitations and to ensure that it is fit for the proposed application [17]. ELISA contains numerous steps that make this very conventional method labor intensive and time consuming. Nonetheless, recording the signal intensities is not the end of this lengthy procedure. Original data that were directly measured by the instrument should not be presented as obtained. Raw data is always the subject of debate.

Therefore raw data should be further processed through a precise data analysis. For instance, while the actual detection results might show a greater performance of one platform over another, the negative controls can draw a completely different conclusion. Negative controls are the result of the assay that has been conducted in the absence of targeted analyte. Negative controls can lead to the cut-off values using Eq. (1.1) [8]:

$$Cut-off\ value = 2 \times average\ (negative\ controls) \tag{1.1}$$

In an ideal condition, the assay that has been carried out with utterly non-infected samples is expected to result in negative detection signal. But in reality, there is always the risk of error that has to be precisely measured and subtracted from the original data. This is important to know that detection signals are considered as positive results if only they are greater than twice of the mean value of the negative controls or so called "cut-off value" [18]. Errors are not just limited to the assay conducted with intentionally negative samples. Often the assay results in negative detection signals even though it was conducted with positive samples containing certain virus concentration. To clarify the point obtained results can be classified into four categories:

1. *True positives (TP)*: Intentionally positive samples that were detected as positives too.
2. *True negatives (TN)*: Intentionally negative samples that were detected as negatives too.
3. *False positives (FP)*: Intentionally negative samples that were detected as positives.
4. *False negatives (FN)*: Intentionally positive samples that were detected as negatives.

These four classes of the results can be used to establish important parameters in evaluation of the assays *such as sensitivity* and *specificity* [13–16]. In that perspective, sensitivity and specificity of the analytical methodology can be calculated by using Eqs. 1.2 and 1.3, respectively [12]:

$$Sensitivity = \frac{TP}{TP + FN} \tag{1.2}$$

$$Specificity = \frac{TN}{TN + FP} \tag{1.3}$$

As another key factor, *accuracy* of the assay is essential when a new developed analytical system has to be tested. *Accuracy* can be calculated by comparing negative and positive values (true and false) in comparison to the total number of the readings (Eq. 1.4) [12]:

$$Accuracy = \frac{TP + TN}{Total\ number\ of\ conducted\ replicates} \quad (1.4)$$

Every newly developed analytical system should be calibrated by running the assay in different known concentrations of the targeted analyte, which leads to the calibration curves. The squared correlation coefficient (R^2) of the plotted calibration curves is the number that shows how precise the data fit a statistical model [19]. R^2 ranges from 0 to 1; the closer the R^2 value is to 1, the more accurate obtained results are. Therefore calibration is one of the necessary steps in assay evaluation as it reveals the level of precision at which the analytical system operates.

Limit of detection (LoD) is lowest quantity of the targeted analyte that developed bioanalytical system can likely determine within a stated confidence limit. It is, therefore, expected to be greater than *limit of blank (LoB)* [20]. This particular parameter has drawn a great deal of importance especially when early detection is concerned. There exist different methods for calculation of *LoD*. In a typical assay evaluation, *LoD* can be determined following Eq. 1.5 [21]:

$$LoD = \frac{3 \times Standard\ deviation}{Slope\ of\ the\ calibration\ curve} \quad (1.5)$$

This is important to note that given equation is valid for *LoD* calculation only when the number of positive replicates used in the assay is greater than negatives samples.

1.5   Advantages and Disadvantages of ELISA

While ELISA benefits from several aspects such as cost effectiveness, safety and versatility of the assay, it is known for considerable disadvantages that have been frequently reported. Labor intensive ELISA has proven to be lengthy and relatively inconsistent with large errors in reproducing the results [8, 9]. It also requires considerably large sample volume for more accurate biorecognition [8, 9]. Possibly one of the major drawbacks of ELISA well plates that has been relatively ignored is the poor performance of the ELISA substrates used in fabrication of such analytical kits. Polystyrene (PS) and/or poly methyl methacrylate (PMMA) are typical materials of choice for mass production of ELISA kits as they offer unique properties such as cost-effectiveness, low specific weight, high impact resistance and flexibility. Nonetheless both polymers are inert in their nature and generally do not contain reactive surface functionalities to promote bimolecular interaction [9, 13, 15, 22–27]. Absence of active functional groups such as amine ($-NH_2$), carboxyl ($-COOH$), hydroxyl ($-OH$) or sulfhydryl ($-SH$) yields in lower performance of such polymeric platforms [28, 29]. Therefore, presence of surface reactive functionalities on the supporting substrates could lead to an efficient analyte-surface interaction, particularly, when the covalent immobilization is aimed [9, 26].

Representative generated surface functional groups

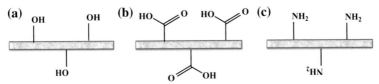

Fig. 1.2 Different surface functional groups that can be generated via modification techniques: **a** hydroxyl; **b** carboxyl; and **c** amine groups

1.6 Functionalization Techniques

To improve current analytical systems, many efforts have been made in order to generate desirable functional groups on the surface of the detection platforms. In that path, commonly applied modification techniques were employed such as plasma treatment [30–37], ultra violet (UV) treatment [38–41] and wet-chemical treatment [42–45] for generation of different functional groups. Figure 1.2 presents different functionalities that might be generated on the surface of the modified samples.

Nonetheless, these modification techniques have shown to have drawbacks that often make their applications limited. For example, the toxicity of the waste materials used in wet chemical treatment makes this technique less eco-friendly [46]. In the UV treatment there is a poor control over concentration and type of the generated functionalities on the modified platforms [46]. Even if desirable functional groups were successfully produces on the bioreceptor's surface, they would lose their activities after a certain period of time when plasma treatment is the applied technique [9]. Functionalized platforms become deactivated over the time since such surfaces have proven to have a substantial tendency to return to their original level of energy [9]. Created surface functional groups, after a brief period of time, reorient themselves in order to occupy lower states of energy. This phenomenon is known as "aging effect" that highly affects the efficiency of activated platforms as it weakens the shelf-life of the activated bioreceptor surfaces.

In the surface activation, what matters even more than generation of desirable functional groups is the ideal concentration and even distribution of such functionalities on the surface. A successful biomolecule immobilization to the modified surface requires a favorable foundation to accommodate protein in an efficient manner. Figure 1.3 depicts a schematic representation of protein behavior towards functionalized surfaces. As it can be observed from Fig. 1.3a, insufficient number of functionalities my lead the proteins to *fall* on the surface and subsequently lose their activity. Such biomolecular entities are highly sensitive towards solid surfaces and can easily lose their activities in a close proximity to the substrate [13]. On the contrary, a highly concentrated surface with functional groups might not necessarily be capable of efficient accommodation of the proteins on the surface as well. Overly functionalized bioreceptor platforms can, in turn, make the protein immobilization

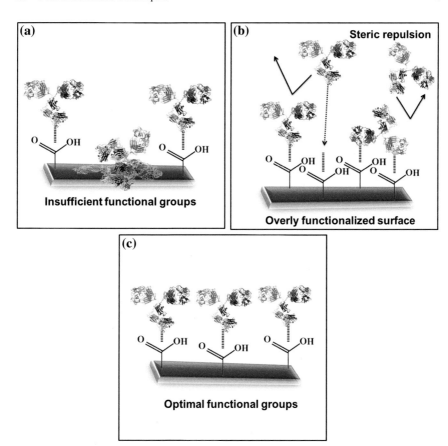

Fig. 1.3 Proteins' responses towards functionalized surfaces: **a** loss of protein activity in the case of insufficiently generated surface functional groups; **b** protein deactivation in the case of overly functionalized surface; and **c** efficient protein immobilization in the case of optimally functionalized surface [12]

troublesome due to the steric repulsion [13, 15]. Moreover, in an overly functionalized platform, multiple binding between functional groups and individual biomolecules can also be the case that can lead to the deactivation of the surface (Fig. 1.3b). Optimal concentration of the surface functional groups, conversely, permits the biomolecules to encounter the surface and its functionalities and consequently bind them in an effective manner. Therefore, it can be concluded that a favorable bioreceptor platform is credited not just for the presence of desirable surface functional groups but also for the optimum concentration of such functionalities on the surface.

In that perspective and in order to response to the mentioned factual requirements for a well-designed bioreceptor, numerous efforts have been made to produce a protein-friendly platform with stable and optimum surface functionalities associated with higher sensitivity and specificity. In Chap. 2, a novel outlook for design and fabrication of novel polymeric platforms is described that can successfully address mentioned drawbacks while amplifying the performance of the commonly applied clinical method, ELISA.

References

1. Li XZB, Li W, Lei X, Fan X, Tian L, Zhang H, Zhang Q (2014) Preparation and characterization of bovine serum albumin surface-imprinted thermosensitive magnetic polymer microsphere and its application for protein recognition. Biosens Bioelectron 51:261–267
2. Nie X-M, Huang R, Dong C-X, Tang L-J, Gui R, Jiang J-H (2014) Plasmonic ELISA for the ultrasensitive detection of *Treponema pallidum*. Biosens Bioelectron 58:314–319. doi:10. 1016/j.bios.2014.03.007
3. Lin T-W, Kekuda D, Chu C-W (2010) Label-free detection of DNA using novel organic-based electrolyte-insulator-semiconductor. Biosens Bioelectron 25(12):2706–2710. doi:10.1016/j. bios.2010.04.041
4. Kirsch J, Siltanen C, Zhou Q, Revzin A, Simonian A (2013) Biosensor technology: recent advances in threat agent detection and medicine. Chem Soc Rev 42(22):8733–8768. doi:10. 1039/C3CS60141B
5. Le Goff A, Holzinger M, Cosnier S (2011) Enzymatic biosensors based on SWCNT-conducting polymer electrodes. Analyst 136(7):1279–1287
6. Liu Y, Li CM (2012) Advanced immobilization and amplification for high performance protein chips. Anal Lett 45(2–3):130–155. doi:10.1080/00032719.2011.633187
7. Turner APF (2013) Biosensors: sense and sensibility. Chem Soc Rev 42(8):3184–3196. doi:10.1039/C3CS35528D
8. Alcon S, Talarmin A, Debruyne M, Falconar A, Deubel V, Flamand M (2002) Enzyme-linked immunosorbent assay specific to dengue virus type 1 nonstructural protein NS1 reveals circulation of the antigen in the blood during the acute phase of disease in patients experiencing primary or secondary infections. J Clin Microbiol 40(2):376–381. doi:10.1128/ jcm.40.02.376-381.2002
9. Hosseini S, Ibrahim F, Djordjevic I, Koole LH (2014) Recent advances in surface functionalization techniques on polymethacrylate materials for optical biosensor applications. Analyst 139(12):2933–2943
10. Dähnrich C, Pares A, Caballeria L, Rosemann A, Schlumberger W, Probst C, Mytilinaiou M, Bogdanos D, Vergani D, Stöcker W, Komorowski L (2009) New ELISA for detecting primary biliary cirrhosis–specific antimitochondrial antibodies. Clin Chem 55(5):978–985. doi:10. 1373/clinchem.2008.118299
11. Xu H, Di B, Pan Y-x, Qiu L-w, Wang Y-d, Hao W, He L-j, Yuen K-y, Che X-y (2006) Serotype 1-specific monoclonal antibody-based antigen capture immunoassay for detection of circulating nonstructural protein NS1: implications for early diagnosis and serotyping of dengue virus infections. J Clin Microbiol 44(8):2872–2878. doi:10.1128/jcm.00777-06
12. Hosseini S (2015) Novel polymeric platforms for biosensing applications. LAP Lambert Academic Publishing, Saarbrücken. ISBN 978-973-659-69472-69471
13. Hosseini S, Ibrahim F, Djordjevic I, Rothan HA, Yusof R, van der Mareld C, Koole LH (2014) Synthesis and processing of ELISA polymer substitute: the influence of surface chemistry and morphology on detection sensitivity. Appl Surf Sci 317:630–638. doi:10.1016/j.apsusc.2014. 08.167
14. Hosseini S, Azari P, Farahmand E, Gan SN, Rothan HA, Yusof R, Koole LH, Djordjevic I, Ibrahim F (2015) Polymethacrylate coated electrospun PHB fibers: an exquisite outlook for

fabrication of paper-based biosensors. Biosens Bioelectron 69:257–264. doi:10.1016/j.bios. 2015.02.034

15. Hosseini S, Ibrahim F, Djordjevic I, Rothan HA, Yusof R, van der Marel C, Benzina A, Koole LH (2014) Synthesis and characterization of methacrylic microspheres for biomolecular recognition: ultrasensitive biosensor for dengue virus detection. Eur Polymer J 60:14–21. doi:10.1016/j.eurpolymj.2014.08.010

16. Hosseini S, Ibrahim F, Rothan HA, Yusof R, van der Marel C, Djordjevic I, Koole LH (2015) Aging effect and antibody immobilization on –COOH exposed surfaces designed for dengue virus detection. Biochem Eng J 99:183–192. doi:10.1016/j.bej.2015.04.001

17. Armbruster DA, Pry T (2008) Limit of blank, limit of detection and limit of quantitation. Clin Biochemist Rev 29(Suppl 1):S49–S52

18. Linares EM, Pannuti CS, Kubota LT, Thalhammer S (2013) Immunospot assay based on fluorescent nanoparticles for dengue fever detection. Biosens Bioelectron 41:180–185. doi:10. 1016/j.bios.2012.08.005

19. Glantz SA, Slinker BK (1990) Primer of applied regression and analysis of variance. McGraw-Hill, New York

20. Clinical and Laboratory Standards Institute (2004) Protocols for determination of limits of detection and limits of quantitation, approved guideline. Wayne, PA USA, 24

21. Shrivastava A, Gupta VB (2011) Methods for the determination of limit of detection and limit of quantitation of the analytical methods. Chronicles Young Sci 2(1):21

22. Lu W, Cao X, Tao L, Ge J, Dong J, Qian W (2014) A novel label-free amperometric immunosensor for carcinoembryonic antigen based on Ag nanoparticle decorated infinite coordination polymer fibres. Biosens Bioelectron 57:219–225. doi:10.1016/j.bios.2014.02.027

23. Hong C-C, Chen C-P, Horng J-C, Chen S-Y (2013) Point-of-care protein sensing platform based on immuno-like membrane with molecularly-aligned nanocavities. Biosens Bioelectron 50:425–430. doi:10.1016/j.bios.2013.07.016

24. Chen J-P, Ho K-H, Chiang Y-P, Wu K-W (2009) Fabrication of electrospun poly(methyl methacrylate) nanofibrous membranes by statistical approach for application in enzyme immobilization. J Membr Sci 340(1–2):9–15. doi:10.1016/j.memsci.2009.05.002

25. Tang C, Saquing CD, Sarin PK, Kelly RM, Khan SA (2014) Nanofibrous membranes for single-step immobilization of hyperthermophilic enzymes. J Membr Sci 472:251–260. doi:10. 1016/j.memsci.2014.08.037

26. Hosseini S, Ibrahim F, Djordjevic I, Koole LH (2014) Polymethyl methacrylate-co-methacrylic acid coatings with controllable concentration of surface carboxyl groups: a novel approach in fabrication of polymeric platforms for potential bio-diagnostic devices. Appl Surf Sci 300:43–50. doi:10.1016/j.apsusc.2014.01.203

27. Hosseini S, Ibrahim F, Djordjevic I, Aeinehvand MM, Koole LH (2014) Structural and end-group analysis of synthetic acrylate co-polymers by matrix-assisted laser desorption time-of-flight mass spectrometry: distribution of pendant carboxyl groups. Polym Test 40:273–279

28. Mitchell JS (2011) Spin-coated methacrylic acid copolymer thin films for covalent immobilization of small molecules on surface plasmon resonance substrates. Eur Polym J 47(1):16–23. doi:10.1016/j.eurpolymj.2010.10.028

29. Grama S, Boiko N, Bilyy R, Klyuchivska O, Antonyuk V, Stoika R, Horak D (2014) Novel fluorescent poly(glycidyl methacrylate)—silica microspheres. Eur Polym J 56:92–104. doi:10. 1016/j.eurpolymj.2014.04.011

30. Chan CM, Ko TM, Hiraoka H (1996) Polymer surface modification by plasmas and photons. Surf Sci Rep 24(1–2):1–54. doi:10.1016/0167-5729(96)80003-3

31. Hegemann D, Brunner H, Oehr C (2003) Plasma treatment of polymers for surface and adhesion improvement. Nucl Instr Methods Phys Res, Sect B 208:281–286

32. Kim K, Lee K, Cho K, Park C (2002) Surface modification of polysulfone ultrafiltration membrane by oxygen plasma treatment. J Membr Sci 199(1):135–145

33. Yang J, Bei J, Wang S (2002) Enhanced cell affinity of poly(D, L-lactide) by combining plasma treatment with collagen anchorage. Biomaterials 23(12):2607–2614

34. Lai J, Sunderland B, Xue J, Yan S, Zhao W, Folkard M, Michael BD, Wang Y (2006) Study on hydrophilicity of polymer surfaces improved by plasma treatment. Appl Surf Sci 252 (10):3375–3379
35. Morent R, De Geyter N, Verschuren J, De Clerck K, Kiekens P, Leys C (2008) Non-thermal plasma treatment of textiles. Surf Coat Technol 202(14):3427–3449
36. Liu C-j, Vissokov GP, Jang BWL (2002) Catalyst preparation using plasma technologies. Catal Today 72(3):173–184
37. Ihara T, Miyoshi M, Ando M, Sugihara S, Iriyama Y (2001) Preparation of a visible-light-active TiO_2 photocatalyst by RF plasma treatment. J Mater Sci 36(17):4201–4207
38. Gassan J, Gutowski VS (2000) Effects of corona discharge and UV treatment on the properties of jute-fibre epoxy composites. Compos Sci Technol 60(15):2857–2863
39. Xin J, Daoud W, Kong Y (2004) A new approach to UV-blocking treatment for cotton fabrics. Text Res J 74(2):97–100
40. Li J, Kim J-K, Lung Sham M (2005) Conductive graphite nanoplatelet/epoxy nanocomposites: effects of exfoliation and UV/ozone treatment of graphite. Scripta Mater 53(2):235–240
41. Khan MA, Haque N, Al-Kafi A, Alam M, Abedin M (2006) Jute reinforced polymer composite by gamma radiation: effect of surface treatment with UV radiation. Polym Plast Technol Eng 45(5):607–613
42. Langley LA, Fairbrother DH (2007) Effect of wet chemical treatments on the distribution of surface oxides on carbonaceous materials. Carbon 45(1):47–54
43. Rohr T, Ogletree DF, Svec F, Fréchet JMJ (2003) Surface functionalization of thermoplastic polymers for the fabrication of microfluidic devices by photoinitiated grafting. Adv Funct Mater 13(4):264–270. doi:10.1002/adfm.200304229
44. Ciampi S, Harper JB, Gooding JJ (2010) Wet chemical routes to the assembly of organic monolayers on silicon surfaces via the formation of Si–C bonds: surface preparation, passivation and functionalization. Chem Soc Rev 39(6):2158–2183
45. Cheng Z, Zhou Q, Wang C, Li Q, Wang C, Fang Y (2011) Toward intrinsic graphene surfaces: a systematic study on thermal annealing and wet-chemical treatment of SiO_2-supported graphene devices. Nano Lett 11(2):767–771
46. Goddard JM, Hotchkiss JH (2007) Polymer surface modification for the attachment of bioactive compounds. Prog Polym Sci 32(7):698–725. doi:10.1016/j.progpolymsci.2007.04. 002

Chapter 2
An Alternative Chemical Approach for Development of Polymeric Analytical Platforms

Abstract This chapter introduces a chemical approach as an alternative solution to solve the problems stated in Chap. 1. Chemically synthesized copolymers made from methyl methacrylate (MMA) and methacrylic acid (MAA) monomers in varied molar ratios of the monomers are proposed for fabrication of novel bioreceptor platforms. Our newly developed materials inherit the presence of surface carboxyl groups (–COOH) from the MAA monomers involved in free-radical polymerization reaction. The concentration of surface functional groups can be tuned by variation of the monomers thus a careful control over surface properties can be achieved. This chapter also reviews the principles of applied characterization techniques such as scanning electron microscopy (SEM), atomic force microscopy (AFM), water-in-air contact angle (WCA) measurement and X-ray photoelectron spectroscopy (XPS) that have been used for investigation of the developed platforms. Since developed bioreceptor surfaces are design for their application in virus detection, the chapter covers the major topics such as infectious diseases and neglected tropical diseases (NTDs). Dengue virus (DENV) has been chosen and tested on the developed bioreceptor platforms.

Keywords Carboxyl groups · Surface properties · Free-radical polymerization · Neglected tropical diseases · Dengue virus

2.1 Proposed Chemical Methodology

As a solution to the existing problems discussed in Chap. 1, a copolymeric system is proposed as a substitute material for fabrication of the bioreceptor platforms. Namely methyl methacrylate (MMA, appendix Table 1) and methacrylic acid (MAA, appendix Table 2) have been chosen as monomers for synthesis of the copolymer poly methylmethacrylate-co-methacrylic acid, poly(MMA-co-MAA). Poly methylmethacrylate (PMMA) is one of the widely used materials for mass production of analytical kits. Between the chemical structure of the proposed copolymer and the commercial PMMA, there exists a small alteration that is the

© The Author(s) 2016
S. Hosseini and F. Ibrahim, *Novel Polymeric Biochips for Enhanced Detection of Infectious Diseases*, SpringerBriefs in Forensic and Medical Bioinformatics, DOI 10.1007/978-981-10-0107-9_2

Fig. 2.1 Poly
(MMA-co-MAA) chemical
structure

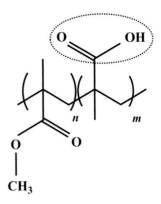

presence of the second monomer (MAA) in the polymerization reaction. Such a minor difference can have a great impact on the performance of the substrate in comparison to the PMMA. As the structure suggests (Fig. 2.1), presence of MAA monomer in preparation of the copolymer poly(MMA-co-MAA) introduces one of the desirable functional groups to the polymeric matrix. With this strategy, instead of applying current modification techniques for generation of such functionalities, a copolymeric material can be obtained, which naturally contains the pendent carboxyl groups (–COOH) inside its chemical structure and at the outmost layer of the surface. Poly(MMA-co-MAA) is polymerized in free-radical polymerization reaction while the surface –COOH groups were derived from MAA monomers of this copolymer [1]. Surface functional groups in the polymeric matrix are part of the chemical structure of the copolymer; hence they would not be deactivated over the time and could not be affected by aging effect or reorientation phenomena [2].

In order to avoid insufficiently or overly functionalized surfaces (Fig. 1.3a, b) and to have a close control over the surface, different molar ratios of the monomers were used in synthesis reaction. As a result different copolymer compositions have been polymerized to have a range of –COOH concentration in the matrix. The variation in monomers' ratio is the key factor to lead to the optimum concentration of functional groups, which can facilitate effective protein immobilization. Table 2.1 presents percentages of each monomer involved in free-radical polymerization reaction. As a control, pure PMMA has also been synthesized under the exact same reaction condition. To simplify the discussion, further in the text, different copolymer compositions are referred as follows: PMMA, comp.(9:1), comp.(7:3), and comp.(5:5). Expectedly, the concentration of –COOH groups in copolymer compositions increases as the molar ratios of the MAA segments increases.

Table 2.1 Different
compositions of poly
(MMA-co-MAA)

Composition	MMA (%)	MAA (%)
PMMA	100	0
Comp.(9:1)	90	10
Comp.(7:3)	70	30
Comp.(5:5)	50	50

2.2 Application of Developed Polymer Compositions in ELISA

Performance of the synthesized copolymer compositions can be assessed by direct application of the materials in the clinical assay. However, it will be a long journey for each discovery to successfully open its way from the laboratory benches to the industry. Polymer compositions developed in our laboratories have obtained the patent (UMCIC Malaysia/PI 2014700658) and are in the process of commercialization. Nevertheless, there is still no available analytical kits made of poly (MMA-co-MAA) for the careful assessment of the method. For that reason, performance of newly developed platforms has been investigated via intermediary substrates that carry prepared copolymer compositions into the assay. Different materials were used as the supporting substrate and subsequently range of behaviors and performances have correspondingly been obtained due to the different morphologies and surface chemistries. Herein, fabrication and processing of polymethacrylate biochips, by using spin coating technique on the silicon substrate, are reported in a great detail [1]. Developed bioreceptor surfaces, in this study, have been chosen as the mediums to introduce synthesized copolymers into the assay. Proposed platforms were thoroughly characterized by different techniques in order to study the surface property and micro-morphology of the samples. After thorough analysis of the samples, performances of the developed surfaces of different compositions in virus detection have been investigated [1–3].

2.3 Characterization of the Biochips

Every newly developed material has to be carefully analyzed. It is also of a great importance to have a better understanding about the characteristics of the coated biochips made from different compositions. Different characterization techniques were used to study the bulk and surface properties of the biochips. In present book, however, the discussion is limited to the surface characterization techniques, as this study assesses the performances of the developed platforms with the strong emphasis on the influence of surface morphology and chemistry on the detection efficiency. The fundamental principles behind applied characterization techniques are as follows.

2.3.1 Scanning Electron Microscopy (SEM), Morphology Analysis

Frontal view and cross-section images of the biochips were recorded by scanning electron microscopy (SEM, JEOL, JSM7600F) [1, 3, 4]. This characterization technique operates with the electron beam that can scan the material and provide information in regard to the morphology of the analyzed samples. Accelerated electrons contain significant amount of kinetic energy. When the electron-substrate interaction occurred, electrons become decelerated and it results in dissipated energy that appears in variety of produced signals. Generated signal includes secondary electrons that are responsible for producing SEM images [5–7]. In this technique, samples are routinely placed on the double sided conductive carbon tape and placed inside the chamber. If the samples are from the family of non-conductive materials, imaging can often be troublesome. For that reason and to avoid surface charging, gold or platinum coating is necessary prior to the imaging [3, 8, 9].

2.3.2 Atomic Force Microscopy (AFM), Topography Analysis

Atomic force microscopy (AFM) is the most commonly used technique from the family of scanning probe microscopies (APM). Such techniques collect images by moving a probe over the surface. As the probe moves, it records the height of the surface. AFM offers a nano-scaled three dimensional (3D) profile of the analyzed surface by measuring the force between the probe and the examined substrate. AFM tip gently travels over the surface, scans the forces between the surface and the probe and sends the signals to the feedback method section, where signals can be translated into the images as the final outcomes of the instrument [10–12]. In this study, the surface topography of platforms, were analyzed by AFM (Ambios, Q scope) in a non-contact mode. Important parameters such as mean roughness (Ra), root mean-square roughness (Rq) and total roughness (Rt) were recorded for all of the developed samples [1, 3].

2.3.3 Water-in-Air Contact Angle Measurement, Surface Wettability

Water-in-air contact angle (WCA) measurement is one of the well-known surface analysis methods to explore the wettability of the examined surface. Detailed study on the behavior of the deposited water droplet towards surface provides valuable information that can be referred to the surface properties of the examined surface. The WCA of 90° is ascribed to the wettability of a neutral surface (Fig. 2.2a). Therefore, examined platforms can be categorized in the group of hydrophobic materials if the WCA is greater than 90° (Fig. 2.2b). Conversely, the WCA of less than 90° refers to the relatively hydrophilic materials (Fig. 2.2c).

Super hydrophilic materials are those that have shown very small WCA (only few degrees), while super hydrophobic material are those for which WCA exceeds

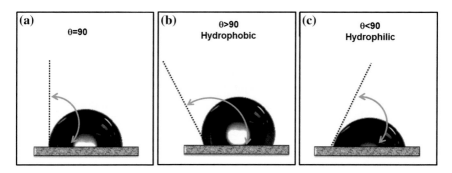

Fig. 2.2 Response of the water droplet deposited on a homogeneous solid surface, which leads to the WCA; **a** neutral; **b** hydrophobic; and **c** hydrophilic

to 150° and above [13, 14]. In this study, WCA measurements were conducted with Dataphysics contact angle system (OCA) by sessile drop method in room temperature by using distilled water (DW). An average contact angle was calculated for five separate measurements with deposited droplets on the center and four corners of the surface. These contact angles were recorded after 1 min of DW deposition while the volume of the droplet was 0.1 μl. Negligible standard deviations have been obtained (±2°) almost for all of the measurements (n = 15, 3 samples from each composition).

2.3.4 X-ray Photoelectron Spectroscopy, Surface Chemistry

X-ray photoelectron spectroscopy (XPS) is a sensitive technique from the class of electron spectroscopy (ES) that can analyze the top 10 nm of the surface. In this technique, a soft X-ray beam (low energy) is irradiated to the sample [15]. The beam excites the electrons from the outmost layer of the surface, which have lower binding energy than the radiated X-ray beam. Excited electrons scape from the parent atom in the form of photoelectron and becomes detected by the instrument [16]. In this study, the XPS measurements were performed by using Quantera SXMtm from Ulvac-PHI (Q1) using monochromatic AlKα-radiation and a take-off angle (θ) of 45° at which the information depth was ∼7 nm. A spot size of 300 × 500 μm was chosen for the sample analysis. Wide-scan measurements were used to identify the presence of elements on the surface. Precise quantification and detailed identification of chemical states was achieved by using narrow-scans. Standard sensitivity parameters were used to convert peak positions to atomic concentrations. Therefore, the concentrations might be deviated from the real values in the absolute sense (relatively less than 20 %) [2].

2.4   Infectious Diseases

Infectious diseases are illnesses caused by pathogenic microorganisms such as viruses, bacteria, parasites or fungi that might live inside or on the body [17]. Such diseases can be widespread through a direct or an indirect way. For example, some infectious diseases can be passed from one person to another while some others are transmitted by bites from insects or animals. Infectious diseases, in some cases, might also be acquired by consuming contaminated food or water or even being exposed to the microorganisms present in the environment. Signs and symptoms can be different depending on the microorganism that caused the infection. Nevertheless, fever and fatigue can be considered as the common symptoms of the infections. While slight complaints may refer to the home remedies, more life-threatening cases may necessitate hospitalization.

2.5   Neglected Tropical Diseases

Neglected tropical diseases (NTDs) are varied types of diseases that considered being as one of the major issues in tropical and subtropical regions. World health organization (WHO) has listed 18 most common and concerning illnesses in the category of NTDs, which are divided in four categories: (i) protozoan diseases such as chagas disease, human african trypanosomiasis and leishmaniases; (ii) bacterial diseases such as buruli ulcer, leprosy and trachoma; (iii) helminth diseases such as cysticercosis/taeniasis, dracunculiasis, echinococcosis, foodborne trematodiases, lymphatic filariasis, onchocerciasis, schistosomiasis, soil-transmitted helminthiases and yaws and; (iv) viral diseases such as dengue, chikungunya and rabies. Such wide-spread diseases affect nearly 1.5 billion people in 149 countries around the world, in particular from the poor populations [17]. NTDs typically grow in underprivileged tropical climates while they have been mostly wiped out in other parts of the world with higher living standards and hygiene [18, 19]. Among different NTDs, dengue fever (DF) has been chosen as the targeted viral infection as the research team which conducted this study is located in one of the tropical countries, Malaysia (Fig. 2.3a).

2.6   Dengue Fever

Along similar infections such as yellow fever, west Nile and Japanese encephalitis, DF can be grouped in the family of *Flaviviridae*. It is a mosquito-borne viral infection, which is mainly transmitted from one person to another by the bite of an aedes mosquito (Fig. 2.3b) [20]. DF cannot be transferred directly from one individual to another.

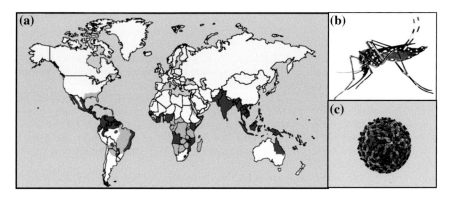

Fig. 2.3 Spreading map of DENV (**a**); aedes mosquito (**b**) dengue enveloped virus (**c**)

According to the WHO's report, the number of the infection with DENV (Fig. 2.3c) exceeds to 50 million cases every year [17] while other sources estimate this number to be up to 100 million infections annually. Only in the first quarter of the year 2014, the outbreak of DF in Malaysia reached to 27,500 cases that consequently resulted in 64 deaths [17]. Although DF is mainly widespread in tropical and subtropical areas, the risk of such infectious illnesses can be worldwide by individuals who contracted the virus while traveling abroad [21]. There are four serotypes for DENV including DENV1, DENV2, DENV3 and DENV4. Infection with any of the serotypes does not immune the person from the infection with another. Although primary infection results in DF, in particular cases, it can develop to the fatal manifestations such as dengue hemorrhagic fever (DHF) and/or dengue shock syndrome (DSS) [21–23]. It is believed that patients who experience DF repeatedly (infection with other serotypes) are more likely to be at risk for syndromes such as DHF and DSS [21–23]. Regular symptoms of DF include fever, nausea, vomiting, headache, skin rashes and pain in the muscles [24–26]. Signs when the patient enters the acute phase of the illness normally are: high fever, restlessness, bleeding under the skin, clammy skin (in the case of DHF) and circulatory collapse (in the case of DSS) [21–23]. With more than half of the world population at risk of this fatal disease, dengue is, indeed, one of the most dangerous mosquito transmitted viral infections [20, 27]. Detection of dengue in the early stages is estimated to reduce the mortality rate from 20 % to below 1 % [24–26].

References

1. Hosseini S, Ibrahim F, Djordjevic I, Koole LH (2014) Polymethyl methacrylate-co-methacrylic acid coatings with controllable concentration of surface carboxyl groups: a novel approach in fabrication of polymeric platforms for potential bio-diagnostic devices. Appl Surf Sci 300:43–50

2. Hosseini S, Ibrahim F, Rothan HA, Yusof R, van der Marel C, Djordjevic I, Koole LH (2015) Aging effect and antibody immobilization on –COOH exposed surfaces designed for dengue virus detection. Biochem Eng J 99:183–192

3. Hosseini S, Ibrahim F, Djordjevic I, Rothan HA, Yusof R, van der Mareld C, Koole LH (2014) Synthesis and processing of ELISA polymer substitute: the influence of surface chemistry and morphology on detection sensitivity. Appl Surf Sci 317:630–638

4. Hosseini S, Ibrahim F, Djordjevic I, Rothan HA, Yusof R, Cvd Marel, Benzina A, Koole LH (2014) Synthesis and characterization of methacrylic microspheres for biomolecular recognition: ultrasensitive biosensor for dengue virus detection. Eur Polymer J 60:14–21

5. Argast A, III Tennis CF (2004) A web resource for the study of alkali feldspars and perthitic textures using light microscopy, scanning electron microscopy and energy dispersive X-ray spectroscopy. J Geosci Educ 52(3):213–217

6. Beane RJ (2004) Using the scanning electron microscope for discovery based learning in undergraduate courses. J Geosci Educ 52(3):250–253

7. Moecher DP (2004) Characterization and identification of mineral unknowns: a mineralogy term project. J Geosci Educ 52(1):5–9

8. Hosseini S, Azari P, Farahmand E, Gan SN, Rothan HA, Yusof R, Koole LH, Djordjevic I, Ibrahim F (2015) Polymethacrylate coated electrospun PHB fibers: an exquisite outlook for fabrication of paper-based biosensors. Biosens Bioelectron 69:257–264

9. Hosseini S, Mahmud NHE, Yahya RB, Ibrahim F, Djordjevic I (2015) Polypyrrole conducting polymer and its application in removal of copper ions from aqueous solution. Mater Lett 149:77–80

10. Lang KM, Hite DA, Simmonds RW, McDermott R, Pappas DP, Martinis JM (2004) Conducting atomic force microscopy for nanoscale tunnel barrier characterization. Rev Sci Instr 75(8):2726–2731

11. Bennig GK (1988) Atomic force microscope and method for imaging surfaces with atomic resolution. Google Patents

12. Binnig G, Quate CF, Gerber C (1986) Atomic force microscope. Phys Rev Lett 56:930–933

13. Yuan Y, Lee TR (2013) Contact angle and wetting properties. In: Bracco G, Holst B (eds) Surface science techniques, vol 51, Springer series in surface sciencesSpringer, Berlin, pp 3–34

14. Grodzka J, Pomianowski A (2006) Wettability versus hydrophilicity. Physicochem Probl Mineral Process 40:5–18

15. Shirley DA (1972) High-resolution X-ray photoemission spectrum of the valence bands of gold. Phys Rev B 5(12):4709–4714

16. Fadley CS (1984) Angle-resolved X-ray photoelectron spectroscopy. Prog Surf Sci 16(3):275–388

17. WHO (2014) World health organization regional office for the Western Pacific

18. Feasey N, Wansbrough-Jones M, Mabey DCW, Solomon AW (2010) Neglected tropical diseases. Br Med Bull 93(1):179–200

19. Yamey G, Hotez P (2007) Neglected tropical diseases. BMJ: Br Med J 335(7614):269–270

20. Bessoff K, Delorey M, Sun W, Hunsperger E (2008) Comparison of two commercially available dengue virus (DENV) NS1 capture enzyme-linked immunosorbent assays using a single clinical sample for diagnosis of acute DENV infection. Clin Vaccine Immunol 15 (10):1513–1518

21. Shu P-Y, Chen L-K, Chang S-F, Yueh Y-Y, Chow L, Chien L-J, Chin C, Lin T-H, Huang J-H (2003) Comparison of capture immunoglobulin M (IgM) and IgG enzyme-linked immunosorbent assay (ELISA) and nonstructural protein NS1 serotype-specific IgG ELISA for differentiation of primary and secondary dengue virus infections. Clin Diagn Lab Immunol 10(4):622–630

22. Stevens AJ, Gahan ME, Mahalingam S, Keller PA (2009) The medicinal chemistry of dengue fever. J Med Chem 52(24):7911–7926

23. Alcon S, Talarmin A, Debruyne M, Falconar A, Deubel V, Flamand M (2002) Enzyme-linked immunosorbent assay specific to dengue virus type 1 nonstructural protein NS1 reveals

circulation of the antigen in the blood during the acute phase of disease in patients experiencing primary or secondary infections. J Clin Microbiol 40(2):376–381

24. Allwinn R (2011) Significant increase in travel-associated dengue fever in Germany. Med Microbiol Immunol 200(3):155–159
25. Lapphra K, Sangcharaswichai A, Chokephaibulkit K, Tiengrim S, Piriyakarnsakul W, Chakorn T, Yoksan S, Wattanamongkolsil L, Thamlikitkul V (2008) Evaluation of an NS1 antigen detection for diagnosis of acute dengue infection in patients with acute febrile illness. Diagn Microbiol Infect Dis 60(4):387–391
26. Linares EM, Pannuti CS, Kubota LT, Thalhammer S (2013) Immunospot assay based on fluorescent nanoparticles for dengue fever detection. Biosens Bioelectron 41:180–185
27. Gubler DJ (2002) Epidemic dengue/dengue hemorrhagic fever as a public health, social and economic problem in the 21st century. Trends Microbiol 10(2):100–103

SpringerBriefs in Applied Sciences and Technology

Thermal Engineering and Applied Science

Series editor

Francis A. Kulacki, Minneapolis, MN, USA

More information about this series at http://www.springer.com/series/8884

David Goluskin

Internally Heated Convection and Rayleigh-Bénard Convection

 Springer

David Goluskin
Mathematics Department and
    Center for the Study of Complex Systems
University of Michigan
Ann Arbor, MI, USA

ISSN 2191-530X   ISSN 2191-5318   (electronic)
SpringerBriefs in Applied Sciences and Technology
ISBN 978-3-319-23939-2   ISBN 978-3-319-23941-5   (eBook)
DOI 10.1007/978-3-319-23941-5

Library of Congress Control Number: 2015950838

Springer Cham Heidelberg New York Dordrecht London

Printed on acid-free paper

Springer International Publishing AG Switzerland is part of Springer Science+Business Media (www.
springer.com)

Preface

The purpose of this SpringerBrief is to review heat transfer in layers of convective fluid. Six different configurations are considered—three that are versions of Rayleigh–Bénard (RB) convection, which is driven by differential heating at the boundaries, and three that are driven by uniform internal heating. The essential features of all six models are derived mathematically. The experimental literature is reviewed in depth for the models of internally heated (IH) convection, which are much less studied than their RB counterparts. Experiments on RB convection are treated in less depth, as they have been thoroughly reviewed elsewhere.

Along with placing the various convective models within a conceptual framework that brings out their similarities, we give some minor results not published elsewhere. For instance, a few of the linear instability and energy stability thresholds given in Tables 2.2 and 2.3 either have not been reported before or have been reported with less precision. One of the bounds proven in Sect. 2.3 is also new, as are the visualizations of simulations included as Fig. 1.2.

Chapter 1 provides background and then defines the six configurations under study, the governing equations of our models, and their basic features. Chapter 2 presents results that can be derived mathematically from the governing equations: linear and nonlinear stability thresholds of static states, along with proven bounds on mean temperatures and heat fluxes. For the IH cases only, Chap. 3 gives a quantitative survey of heat transport in both laboratory experiments and numerical simulations, followed by suggestions for future work.

The author is grateful to Charles R. Doering, Erwin P. van der Poel, Jared P. Whitehead, and Francis A. Kulacki for their many helpful comments on the manuscript. Thanks also to Martin Wörner for providing his original data and to Francis A. Kulacki for providing not only his original data but also many hard-to-find references. More general thanks are due to Edward A. Spiegel, who taught the author much of what he knows about convection and many other topics.

Ann Arbor, MI, USA   David Goluskin

Contents

Chapter 1
A Family of Convective Models

Abstract Six canonical models of convection are described: three configurations of internally heated (IH) convection driven by constant and uniform volumetric heating, and three configurations of Rayleigh–Bénard (RB) convection driven by the boundary conditions. The IH models are distinguished by differing pairs of thermal boundary conditions: top and bottom boundaries of equal temperature, an insulating bottom with heat flux fixed at the top, and an insulating bottom with temperature fixed at the top. The RB models too are distinguished by whether temperatures or heat fluxes are fixed at the top and bottom boundaries. Integral quantities important to heat transport are examined, including the mean fluid temperature, the mean temperature difference between the boundaries, and the mean convective heat flux. Integral relations and bounds are presented, and further bounds are conjectured for the IH cases. Similarities and differences between the six configurations are emphasized.

This SpringerBrief is concerned entirely with convection—fluid motion driven by differential body forces. We focus on several simple configurations that lend themselves to theoretical and experimental study. Convection arises in many contexts and figures prominently in astrophysics, geophysics, and certain engineering applications. Astrophysical occurrences include stellar interiors [15, 17, 29, 45] and planetary atmospheres and interiors [22, 26, 28, 41], while terrestrial occurrences include the Earth's outer core [4, 6, 7, 14], mantle [39], oceans [33, 46], and atmosphere [13]. Many of these systems have internal sources or sinks of buoyancy, including the Earth's mantle, the cores of large main-sequence stars, radiating atmospheres, and nearly any engineered system where chemical or nuclear reactions take place in a fluid environment. Among such engineering applications, particular attention has been paid to nuclear accident scenarios in which exothermic nuclear reactions drive convection in molten material [3, 21, 34].

We speak here in terms of thermal convection, where the body forces are gravitational and depend on the fluid's density, which, in turn, depends on its temperature. Other types of convection are not discussed but are often governed by similar dynamics. In compositional convection, for instance, chemical concentration takes the place of temperature. In electroconvection (e.g., [5, 47]), electric charge takes

© Springer International Publishing Switzerland 2016

D. Goluskin, *Internally Heated Convection and Rayleigh-Bénard Convection*,
SpringerBriefs in Applied Sciences and Technology,
DOI 10.1007/978-3-319-23941-5_1

the place of temperature, and electrical potential takes the place of gravitational potential.

Convection can be indefinitely sustained in each configuration studied here, and we focus on the time-averaged properties of sustained convection, especially heat transport. Transient phenomena are not addressed. The minimum requirement for ordinary fluid to convect is that warmer (less dense) fluid lie below cooler (more dense) fluid, and that this adverse temperature gradient be sufficiently destabilizing to overcome the viscous forces that damp fluid motion. For the convection to be indefinitely sustained against viscous dissipation, there is an additional requirement: an inexhaustible source of energy that drives the system away from equilibrium by endlessly adding heat somewhere other than the top of the domain or removing heat somewhere other than the bottom. This can be accomplished through the thermal boundary conditions, as when a pot of water is boiled on a stove, or it can be accomplished through internal heat sources or sinks, as when radioactive decay heats the Earth's mantle.

The present chapter lays out the basic features of six convective configurations— three that are driven solely by the boundary conditions, and three that are driven by internal heating. The configurations are defined in Sect. 1.1, and the Boussinesq equations that are used to model them are given in Sect. 1.2, followed in Sect. 1.3 by our chosen nondimensionalization. The basic commonalities and differences of the six configurations are then summarized: Sect. 1.4 discusses static states, Sect. 1.5 gives a qualitative look ahead to the experimental findings that are surveyed in Chap. 3, and Sect. 1.6 introduces the integral quantities and relations that govern heat transport.

1.1   Six Configurations

The six configurations we study here share the same basic geometry: a horizontal fluid layer of height d. In its horizontal dimensions, the layer can be modeled as infinite, periodic, or bounded, though only the last case is realizable in physical experiments. In a theoretical investigation, we can allow the convection to have three-dimensional (3D) freedom, or we can make it two-dimensional (2D) by imposing uniformity in one of the horizontal dimensions.

The six configurations differ only in their top and bottom thermal boundary conditions and in the presence or absence of volumetric heating. Three of the cases are versions of Rayleigh–Bénard (RB) convection, wherein the flow is driven solely by thermal boundary conditions that cause heat to enter across the bottom boundary and exit across the top one. The other three cases are instances of internally heated (IH) convection. Here, heat is added only by a constant, uniform source, and at least some of it exits across the top boundary. The only thermal boundary conditions we employ are fixed-temperature, fixed-flux, or perfectly insulating. Fixed temperatures model perfectly conductive boundaries, while fixed heat fluxes model boundaries that conduct heat poorly [43].

RB1	RB2	RB3
$T = 0$	$\frac{\partial T}{\partial z} = -\Gamma$	$T = 0$
$T = \Delta$	$\frac{\partial T}{\partial z} = -\Gamma$	$\frac{\partial T}{\partial z} = -\Gamma$

IH1	IH2	IH3
$T = 0$	$\frac{\partial T}{\partial z} = -\Gamma$	$T = 0$
H	H	H
$T = 0$	$\frac{\partial T}{\partial z} = 0$	$\frac{\partial T}{\partial z} = 0$

Fig. 1.1 Schematics of the six configurations studied in the present work. They are distinguished by their thermal boundary conditions and by the presence or absence of a constant and uniform internal heat source (H). Gravity acts vertically downward. All quantities are dimensional; nondimensionalized versions of these schematics appear in Tables 1.1 and 1.2. In the IH2 configuration, H and Γ are related such that heat loss balances internal heat production (see text)

The thermal boundary conditions of our three RB configurations are shown in the top row of Fig. 1.1. In the case we call RB1, which is the most-studied RB model, the boundary temperatures are fixed, with a temperature drop of Δ from the bottom boundary to the top one. (The temperature at the top boundary can be fixed at zero for convenience because, under the governing equations of our models, only temperature *differences* affect the dynamics.) In RB2, the heat flux across each boundary is fixed by setting the temperature gradients there to the same value, $-\Gamma$. The RB3 case mixes the two previous cases, having a fixed-temperature condition on the top boundary and a fixed-flux condition on the bottom one. The configuration where these two boundary conditions are swapped need not be considered since it is related to RB3 by symmetry, at least with the governing equations of our models.

The thermal boundary conditions of our three IH configurations are shown in the bottom row of Fig. 1.1. Temperature is produced volumetrically at rate H in all three cases, corresponding to heat production at rate H/c_p, where c_p is the fluid's heat capacity. In IH1, the top and bottom temperatures are fixed at the same value, and the internally produced heat escapes across both boundaries. In IH2 and IH3, the bottom boundary is perfectly insulating, meaning the vertical temperature gradient must vanish there. The top boundary, across which all the internally produced heat escapes, has a fixed flux in IH2 and a fixed temperature in IH3. In IH2, the boundary flux must match the rate of internal heat production in order for convection to be statistically steady, so Γ is determined by H, as described in Sect. 1.3. Configurations where the boundary conditions in IH2 or

IH3 are reversed, making the tops insulating, do not need to be considered; fluid in such configurations would remain static, as follows from the stability results of Sect. 2.2. Notice that the temperature is subject to Dirichlet conditions in RB1 and IH1, Neumann conditions in RB2 and IH2, and one condition of each type in RB3 and IH3.

For the velocity, we impose either no-slip or free-slip boundary conditions. No-slip conditions apply to most laboratory experiments and many engineering applications, while free-slip conditions are more appropriate in modeling certain astrophysical, geophysical, and plasma physical systems. Although we address both possibilities, some results are available only for no-slip boundaries. If side boundaries exist, we assume they are perfect thermal insulators.

The six models of Fig. 1.1 can be extended in various ways. For instance, internal heating can be added to the RB configurations, creating hybrid models driven both by the boundary conditions and by internal heating. The thermal boundary conditions can also be made more complicated, perhaps to model thermal radiation or moderate conductivity. However, such models require at least one more control parameter than those of Fig. 1.1. When modeled by the Boussinesq equations, which are introduced in the next section, each of the six configurations we study is governed by *only two* dimensionless control parameters, aside from any parameters used to describe the geometry. In a sense described at the end of Sect. 1.3, they are the only models of convective layers for which this is true, hence they are a natural starting point.

In many sections of this SpringerBrief results are presented for all six configurations in Fig. 1.1, letting us highlight their similarities and differences. Section 2.3 and Chap. 3, however, focus mainly on IH convection. This work is not meant to be a comprehensive review of RB convection, which has been reviewed several times in recent decades [1, 11, 18, 32, 40]. Interest in RB convection goes well beyond heat transport, as the system has become a canonical model of nonlinear science, having provided early examples of instabilities, bifurcations, pattern formation, and chaos in spatially extended systems. IH convection, which has been the subject of numerous works but is still much less studied than its RB counterpart, was last reviewed in the 1980s [10, 31]. Bringing this topic up to date requires not only that we review the more recent studies of IH convection, which are relatively few in number, but also that we reinterpret older studies in light of our contemporary understanding of RB convection.

Many of the laboratory experiments reviewed in Chap. 3 are captured well by one of our three IH models, but applications are rife with further complications, including compressibility, temperature-dependent material properties, complicated geometries, chemical and nuclear reactions, rotation, magnetism, and other thermal boundary conditions. Nonetheless, characterizing heat transport in the relatively simple models we consider is already a formidable challenge, and the task is far from complete.

1.2 Boussinesq Equations

A mathematical model of thermal convection must include equations governing the velocity and temperature fields, along with a constitutive relation between temperature and density. The compressible Navier–Stokes equations [50] describe the velocity field accurately in a wide range of physical applications, but they are challenging equations to study analytically or integrate numerically, and they can be avoided when the density field does not deviate too strongly from hydrostatic equilibrium. Pressure-driven flows with weak density variations can simply be approximated as having constant densities, yielding the incompressible Navier–Stokes equations. In buoyancy-driven flows, however, density variations cannot be totally ignored because they create the buoyancy gradients needed to drive motion. The typical compromise is to employ the Boussinesq approximation, as we do here.

The Boussinesq approximation, which was first invoked by Oberbeck [35] and is also called the Oberbeck–Boussinesq approximation, involves two main assumptions. First, the fluid's density, ρ, is assumed to vary linearly with temperature, T, about some hydrostatic reference state denoted by ρ_* and T_*. That is,

$$\rho(T) = \rho_* \left[1 - \alpha(T - T_*)\right], \tag{1.1}$$

where α is the linear coefficient of thermal expansion. Second, the density variations are assumed to be sufficiently weak that they can be ignored everywhere except in the buoyancy force. The fluid is sometimes called incompressible since the velocity field is divergence-free, although compressibility does manifest in the buoyancy variations. Numerous justifications have been put forth for replacing the fully compressible Navier–Stokes equations with the simpler Boussinesq equations, typically invoking some combination of asymptotic expansions and ad hoc assumptions. See Spiegel and Veronis [44] for one physical justification and Rajagopal et al. [37] for a discussion of various other justifications. The precise assumptions invoked vary, but in all versions there is a sense in which gradients of the fluid's properties should not be too steep. If the Boussinesq approximation is used in modeling a physical system, the assumptions under which the approximation holds should be checked if possible, either by physical measurement or by numerical simulation of compressible equations.

With constant gravitational acceleration g acting in the $-\hat{\mathbf{z}}$ direction, applying the Boussinesq approximation to the compressible Navier–Stokes equations yields the Boussinesq equations [8, 38],

$$\nabla \cdot \mathbf{u} = 0 \tag{1.2}$$

$$\partial_t \mathbf{u} + \mathbf{u} \cdot \nabla \mathbf{u} = -\frac{1}{\rho_*}\nabla p + \nu \nabla^2 \mathbf{u} + g\alpha T \hat{\mathbf{z}} \tag{1.3}$$

$$\partial_t T + \mathbf{u} \cdot \nabla T = \kappa \nabla^2 T + H, \tag{1.4}$$

where $\mathbf{u} = (u, v, w)$ is the fluid's velocity vector, p its pressure, ν its kinematic viscosity, and κ its thermal diffusivity. The temperature source term, H, is absent from RB convection but drives the convection in our IH models. The pressure term in (1.3) has absorbed hydrostatic terms of the buoyancy force coming from (1.1).

1.3   Nondimensionalization

To nondimensionalize the Boussinesq equations, we scale distance by the layer height, d, time by the characteristic timescale of thermal diffusion, d^2/κ, and pressure by $\rho_* d^2/\kappa$. We scale temperature by a dimensional quantity, Δ, that is defined differently in various configurations. In the RB1 case, Δ is the prescribed temperature difference between the boundaries. In the other cases,

$$\Delta := \begin{cases} d\Gamma & \text{RB2, RB3, IH2} \\ \frac{d^2 H}{\kappa} & \text{IH1, IH2, IH3.} \end{cases} \tag{1.5}$$

Nondimensionalized by these Δ, the temperature difference between the boundaries is unity in RB1; the fixed temperature fluxes are unity in RB2, RB3, and IH2; and the volumetric heating rate is unity in the IH cases. Both definitions in (1.5) apply to IH2 because we add the consistency condition $\Gamma = dH/\kappa$ in that case to ensure that heat production balances heat loss.

The Boussinesq equations (1.2)–(1.4) in dimensionless form are

$$\nabla \cdot \mathbf{u} = 0 \tag{1.6}$$

$$\partial_t \mathbf{u} + \mathbf{u} \cdot \nabla \mathbf{u} = -\nabla p + Pr\nabla^2 \mathbf{u} + PrRT\hat{\mathbf{z}} \tag{1.7}$$

$$\partial_t T + \mathbf{u} \cdot \nabla T = \nabla^2 T + Q, \tag{1.8}$$

where the symbols $\mathbf{u}$, T, p, $\mathbf{x}$, and t henceforth represent dimensionless quantities. The vertical extent is $0 \leq z \leq 1$, and the temperature source term is

$$Q = \begin{cases} 0 & \text{RB} \\ 1 & \text{IH.} \end{cases} \tag{1.9}$$

The dimensionless control parameters are the Rayleigh number, R, and Prandtl number, Pr, defined by

$$R := \frac{g\alpha d^3 \Delta}{\kappa \nu} \tag{1.10}$$

$$Pr := \frac{\nu}{\kappa}. \tag{1.11}$$

The definition of R differs between cases when the definition (1.5) of Δ differs.

The Rayleigh number may be thought of as the ratio of inertial forces to viscous forces. When it is large, the fluid is strongly driven by differential buoyancy forces. We regard R as the primary control parameter since raising it typically makes the flow more complex. For R to indeed be a control parameter, we needed to define the dimensional temperature scale, Δ, in terms of quantities that are known a priori: the boundary conditions or heating rate. However, it is sometimes useful to define a different Rayleigh number using a temperature scale that is determined dynamically by the flow. This sort of Rayleigh number cannot serve as a control parameter but can be a useful diagnostic quantity. Thus, we will sometimes distinguish between the *control* Rayleigh number, R, and *diagnostic* Rayleigh numbers, Ra or $\widetilde{Ra}$, defined in Sect. 1.6.5.

The Prandtl number is the rate at which the fluid diffuses momentum, relative to the rate at which it diffuses heat. Unlike the Rayleigh number, it is a material property of the fluid and does not depend on the geometry or boundary conditions. The Prandtl number is large in fluids that damp motion strongly and conduct heat poorly. In the Earth's mantle, for instance, Pr is effectively infinite. The Prandtl number is small in fluids that damp motion weakly and conduct heat well, such as liquid metals and stellar plasmas. Air and water are intermediate examples, having Prandtl number close to 0.7 and 7, respectively, under atmospheric conditions.

In all six configurations of Fig. 1.1, the dynamics depend on only two control parameters, Pr and R, aside from any parameters describing the geometry, such as aspect ratios. This is the minimum number of parameters we can hope for in the study of convection, except in those special cases where Pr can be eliminated because it is effectively infinite. Additional parameters would be needed if the thermal boundary conditions were more complicated [24, 43], the internal heating law were more complicated [16, 48], or the boundary conditions and internal heating each introduced their own temperature scales [2, 9, 23, 27, 30, 42, 52]. We restrict ourselves to models that require only Pr and R since every additional parameter makes it much harder to understand parameter space. In fact, among the ways that uniform heating, fixed-temperature boundaries, and fixed-flux boundaries can be combined, our six configurations (and their symmetry-related siblings) seem to be the only ones governed by so few parameters.

1.4 Static States

Dimensionless schematics of the RB and IH configurations are shown in the first rows of Tables 1.1 and 1.2, respectively. The basic features of the six cases are summarized in the subsequent rows of both tables and are further laid out in the remainder of this chapter. The simplest solutions to the governing equations are static, with heat transported only by conduction. These are the unique asymptotic states when R is sufficiently small (cf. Sect. 2.2), and they solve the Poisson or Laplace equation $\nabla^2 T + Q = 0$. Since we assume that side boundaries are nonexistent or perfectly insulating, the static temperature fields, T_{st}, vary only in z:

Table 1.1 Summary of the properties of RB convection discussed in the present chapter

	RB1	RB2	RB3
Configuration	$T = 0$ \\\\\\\\\\ ///////// $T = 1$	$\frac{\partial T}{\partial z} = -1$ \\\\\\\\\\ ///////// $\frac{\partial T}{\partial z} = -1$	$T = 0$ \\\\\\\\\\ ///////// $\frac{\partial T}{\partial z} = -1$
Static temperature profile			
Turbulent temperature profile			
Heat balance	$\left. \frac{d}{dz}\overline{T}\right|_{z=1} = \left. \frac{d}{dz}\overline{T}\right|_{z=0}$		
$\overline{J}(z)$	$1 + \langle wT \rangle$	1	
$\langle J \rangle$	$1 + \langle wT \rangle$	1	
Uniform $\langle wT \rangle$ bounds	$0 \le \langle wT \rangle < \infty$	$0 \le \langle wT \rangle < 1$	
Uniform $\delta\langle T \rangle$ bounds	$0 < \delta\langle T \rangle < 1$	$-\frac{1}{\sqrt{3}} \le \delta\langle T \rangle \le \frac{1}{\sqrt{3}}$	$0 < \delta\langle T \rangle \le \frac{1}{\sqrt{3}}$
Empirical relation	$\delta\langle T \rangle \sim \frac{1}{2}$	$\delta\langle T \rangle \sim \frac{1}{2}\left(1 - \langle wT \rangle\right)$	
N	$1 + \langle wT \rangle$	$\dfrac{1}{1 - \langle wT \rangle}$	

All quantities are dimensionless, and the vertical extent is $0 \le z \le 1$. Notation is defined throughout the chapter. Briefly, $\overline{*}$ denotes an average over horizontal directions and time, $\langle * \rangle$ denotes an average over volume and time, $\delta\langle T \rangle$ is the mean temperature of the fluid relative to that of the top boundary, and $J = -\partial_z T + wT$ is the sum of the conductive and convective vertical heat fluxes

Table 1.2 Summary of the properties of IH convection discussed in the present chapter

	IH1	IH2	IH3
Configuration	$T = 0$ $\\\\\\\\\\\\$ $Q = 1$ $\////////$ $T = 0$	$\frac{\partial T}{\partial z} = -1$ $\\\\\\\\\\\\$ $Q = 1$ $\////////$ $\frac{\partial T}{\partial z} = 0$	$T = 0$ $\\\\\\\\\\\\$ $Q = 1$ $\////////$ $\frac{\partial T}{\partial z} = 0$
Static temperature profile	$\frac{1}{8}$	$\frac{1}{2}$	$\frac{1}{2}$
Turbulent temperature profile			
Heat balance	$-\frac{d}{dz}\overline{T}\big|_{z=1} + \frac{d}{dz}\overline{T}\big|_{z=0} = 1$	$-\frac{d}{dz}\overline{T}\big|_{z=1} = 1$	
$\overline{J}(z)$	$\langle wT \rangle + \left(z - \frac{1}{2}\right)$	z	
$\langle J \rangle$	$\langle wT \rangle$	$\frac{1}{2}$	
Uniform $\langle wT \rangle$ bounds	$0 \le \langle wT \rangle < \frac{1}{2}$	$0 \le \langle wT \rangle < \frac{1}{2} + \frac{1}{\sqrt{3}}$	$0 \le \langle wT \rangle < \frac{1}{2}$
Uniform $\delta\langle T \rangle$ bounds	$0 < \delta\langle T \rangle \le \frac{1}{12}$	$0 < \delta\langle T \rangle \le \frac{1}{3}$	
Empirical relation	$\delta\langle T \rangle \sim \overline{T}_{max}$	$\delta\langle T \rangle \sim \frac{1}{2} - \langle wT \rangle$	
N	$\dfrac{1}{8\overline{T}_{max}}$	$\dfrac{1}{1 - 2\langle wT \rangle}$	
$\widetilde{N}$	$\dfrac{1}{12\,\delta\langle T \rangle}$	$\dfrac{1}{3\,\delta\langle T \rangle}$	

All quantities are dimensionless, and the vertical extent is $0 \le z \le 1$. Notation is defined throughout the chapter. Briefly, $\overline{*}$ denotes an average over horizontal directions and time, $\langle * \rangle$ denotes an average over volume and time, $\delta\langle T \rangle$ is the mean temperature of the fluid relative to that of the top boundary, $J = -\partial_z T + wT$ is the sum of conductive and convective vertical heat fluxes, and $\overline{T}_{max}$ is the maximum value of $\overline{T}(z)$

$$T_{st}(z) = \begin{cases} 1 - z & \text{RB1, RB2, RB3} \\ \frac{1}{2}z(1-z) & \text{IH1} \\ \frac{1}{2}(1-z^2) & \text{IH2, IH3.} \end{cases} \tag{1.12}$$

These purely conductive profiles are depicted in the second rows of Tables 1.1 and 1.2. They are parabolic with internal heating and linear without it. The static profiles in RB2 and IH2 are determined only up to additive constants, but these constants do not affect the dynamics, so we have fixed them for convenience.

In each configuration, our dimensional temperature scale, Δ, is characteristic of the static state. This is why the dimensionless T_{st} have no dependence on R. When we define *diagnostic* Rayleigh number in Sect. 1.6.5, we will do so by replacing Δ with temperature scales characteristic of the convective states, rather than the static ones.

1.5   Temperature Fields in Strong Convection

Whereas the fluid remains static when R is sufficiently small, it convects strongly when R is large. Convection typically strengthens monotonically as R is raised, though this is not universally true and can depend on how strength is quantified. (The non-monotonicity of convective transport in [20] provides a counterexample.) Figure 1.2 shows instantaneous temperature fields from 2D simulations at large R. The RB1 field in Fig. 1.2a is representative of all three RB cases: hot plumes rise from the bottom boundary, cold plumes descend from the top one, and both types of plumes contribute to upward heat transport. In the IH1 field of Fig. 1.2b, cold plumes descend from the top, but the bottom boundary layer is cold and *stably* stratified. This bottom layer emits no buoyant plumes, so any mixing between it and interior must be driven by shear, rather than buoyancy. The IH3 field in Fig. 1.2c is representative of both IH2 and IH3: cold plumes descend from the top boundary, and there is no thermal boundary layer at the bottom.

The turbulent convection that occurs at large R creates mean vertical temperature profiles very different from the static ones. Rough schematics of such profiles are shown in the third rows of Tables 1.1 and 1.2. Although these schematics include no secondary details, they illustrate the main differences between the various configurations. In each of the six cases, mixing by strong convection tends to flatten the temperature profile in the layer's interior. (We are not aware of a counterexample in 3D, though one exists in 2D under conditions that allow very strong winds to develop [20, 51].) The roughly isothermal interiors are flanked by one or two thermal boundary layers, and these are what distinguish the various cases.

In all six of our models, temperature is unstably stratified at the top boundary. At the bottom boundary, the temperature is unstably stratified in the RB cases, stably stratified in IH1, and unstratified in IH2 and IH3. These facts are evident in the static temperature profiles (cf. Tables 1.1 and 1.2) and remain provably true of sustained convection, at least in a time-averaged sense. As convective heat transport rises, the

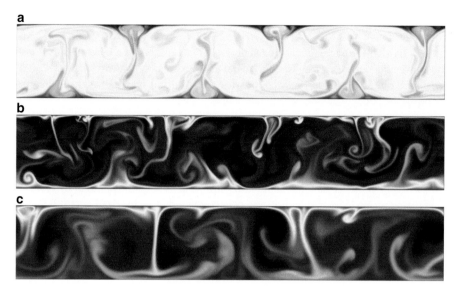

Fig. 1.2 Temperature fields from 2D simulations of (**a**) RB1, (**b**) IH1, and (**c**) IH3. Each simulation employed a horizontal period of 6, no-slip boundaries above and below, $Pr = 1$, and $R/R_L = 10^5$, where R_L is the Rayleigh number at which the static becomes linearly unstable (cf. Sect. 2.1). The coolest fluid (*blue*) has a temperature of zero in each case, and the warmest fluid (*red*) has a temperature of 1, 0.017, and 0.044, respectively

mean temperature profiles undergo various changes. In RB1, where the temperature difference between the boundaries cannot change, the boundary layers steepen as the heat flux through the domain rises. In the other two RB cases, where the mean flux through the domain cannot change, the temperature difference between the boundaries decreases. In the IH cases, the temperature of the fluid, relative to that of the top boundary, drops as the convection strengthens. The produced heat leaves only across the top boundary in IH2 and IH3. It leaves across both boundaries in IH1, but the majority leaves across the top boundary, hence the top boundary layer is steeper than the bottom one.

Although the turbulent temperature profiles are fairly easy to understand qualitatively, it is very difficult in general to anticipate their quantitative features. This would be tantamount to accomplishing our main objective of characterizing the bulk heat transport.

1.6   Mean Heat Fluxes and Integral Relations

Heat in a convecting fluid is transported by two mechanisms simultaneously: conduction and convection. Conduction refers to the diffusion of heat down the temperature gradient, while convection refers to the advection of heat by fluid motion. The temperature equation (1.8) can be written in the standard form of a conservation law as $\partial_t T + \nabla \cdot \mathbf{J} = Q$, where $\mathbf{J} := \mathbf{u}T - \nabla T$ is the total *heat current* at a point. Evidently, the conductive current is $-\nabla T$ in our nondimensionalization, and the convective current is $\mathbf{u}T$. The horizontal components of $\mathbf{J}$ vanish when averaged over the volume since our side boundaries are insulting or nonexistent. Here we focus on the heat current's vertical component, J, where

$$J := J_{\text{cond}} + J_{\text{conv}} \tag{1.13}$$

$$:= -\partial_z T + wT. \tag{1.14}$$

Much of our effort is devoted to quantifying the relative contributions to vertical heat transport made by conduction and convection—that is, by $-\partial_z T$ and wT.

In our notation, an overbar, as in $\bar{f}$, denotes an average over horizontal surfaces and infinite time. Angular brackets, as in $\langle f \rangle$, denote an average over the entire volume and infinite time. When the dimensionless domain is bounded horizontally by $0 \le x \le L_x$ and $0 \le y \le L_y$,

$$\bar{f}(z) := \lim_{\tau \to \infty} \frac{1}{\tau} \frac{1}{L_x L_y} \int_0^\tau dt \int_0^{L_y} dy \int_0^{L_x} dx f(\mathbf{x}, t), \tag{1.15}$$

$$\langle f \rangle := \lim_{\tau \to \infty} \frac{1}{\tau} \frac{1}{L_x L_y} \int_0^\tau dt \int_0^1 dz \int_0^{L_y} dy \int_0^{L_x} dx f(\mathbf{x}, t). \tag{1.16}$$

The above limits can be replaced with lim inf or lim sup to ensure their existence. For simplicity, we assume in our calculations that infinite-time averages commute with vertical averages and that horizontal averages commute with vertical derivatives, though these assumptions can often be avoided. In the above notation, the mean heat flux across a horizontal surface is

$$\bar{J}(z) = -\bar{T}'(z) + \overline{wT}(z), \tag{1.17}$$

where the prime indicates differentiation in z. The mean vertical flux across the entire layer is

$$\langle J \rangle = \delta \bar{T} + \langle wT \rangle, \tag{1.18}$$

where the mean conductive flux

$$\delta \bar{T} := \bar{T}_B - \bar{T}_T \tag{1.19}$$

is the difference between the mean bottom temperature, $\bar{T}_B$, and mean top temperature, $\bar{T}_T$. Expressions (1.17) and (1.18) are simply averages of the definition (1.14) of J; configuration-specific constraints on $\bar{J}(z)$ and $\langle J \rangle$ are given in Sect. 1.6.2.

1.6.1   Heat Balances

Conservation of heat energy is expressed in the various cases by the heat balances

$$\overline{T}'_T = \overline{T}'_B \qquad\qquad \text{RB1, RB2, RB3} \qquad\qquad (1.20)$$

$$-\overline{T}'_T + \overline{T}'_B = 1 \qquad\qquad \text{IH1} \qquad\qquad (1.21)$$

$$-\overline{T}'_T = 1 \qquad\qquad \text{IH2, IH3.} \qquad\qquad (1.22)$$

These balances, which are shown also in the fourth rows of Tables 1.1 and 1.2, are derived by averaging the temperature equation (1.8) over volume and time. Time derivatives vanish from such averages because the instantaneous volume averages of $\mathbf{u}$ and T are bounded uniformly in time (cf. Sect. 2.3). In the RB cases, the balance reflects the fact that mean heat flux into the layer at the bottom $(-\overline{T}'_B)$ must equal the mean flux out of the layer at the top $(-\overline{T}'_T)$. In IH1, the rate of internal heat production (unity) is balanced by the combined outward fluxes of heat across the top boundary $(-\overline{T}'_T)$ and the bottom one $(\overline{T}'_B)$. In IH2 and IH3, where the bottom boundary is insulating, the internal production is balanced entirely by the outward flux across the top boundary $(-\overline{T}'_T)$.

1.6.2   Constraints on Net Heat Fluxes

Little can be said a priori about the variation with height of the mean heat flux components, $-\overline{T}'(z)$ and $\overline{wT}(z)$, but we can derive constraints on their sum, $\bar{J}(z)$. Averaging the temperature equation (1.8) horizontally, vertically from 0 to z, and temporally gives

$$\bar{J}(z) := -\overline{T}'(z) + \overline{wT}(z) = \begin{cases} -\overline{T}'_B & \text{RB1} \\ 1 & \text{RB2, RB3} \\ -\overline{T}'_B + z & \text{IH1} \\ z & \text{IH2, IH3.} \end{cases} \qquad (1.23)$$

The net vertical flux is the same at every height in the RB cases and increases linearly with height in the IH cases. In the four cases where a boundary flux is fixed, $\bar{J}(z)$ is known exactly. In RB1 and IH1, $\bar{J}(z)$ is determined only up to the mean heat flux at the bottom boundary, $-\overline{T}'_B$. Below we give some alternate expressions for $\bar{J}(z)$ in these two cases, but they all involve quantities that, like $-\overline{T}'_B$, are not known a priori.

The volume-averaged heat flux, $\langle J \rangle$, is also constrained. In the four cases where $\bar{J}(z)$ is known exactly, $\langle J \rangle$ is found by vertically integrating (1.23). In RB1 and IH1, the fixed-temperature boundary conditions ensure that the mean conductive fluxes are $\delta\overline{T} = 1$ and $\delta\overline{T} = 0$, respectively. The results are

$$\langle J\rangle := \delta\overline{T} + \langle wT\rangle = \begin{cases} 1+\langle wT\rangle & \text{RB1} \\ 1 & \text{RB2, RB3} \\ \langle wT\rangle & \text{IH1} \\ \frac{1}{2} & \text{IH2, IH3.} \end{cases} \tag{1.24}$$

In all six cases, we would like to know how the control parameters affect the convective flux, $\langle wT\rangle$. In RB1 and IH1, where $\delta\overline{T}$ is fixed, this is equivalent to knowing $\langle J\rangle$. In the other four cases, where $\langle J\rangle$ is fixed, it is equivalent to knowing $\delta\overline{T}$.

In the RB1 and IH1 cases, the expressions (1.23) for $\overline{J}(z)$ can be rewritten by replacing $\overline{T}'_B$ with other integral quantities. Relations between $\overline{T}'_B$ and $\overline{T}'_T$ are provided by the heat balances of Sect. 1.6.1, while relations between $\overline{T}'_B$ and $\langle wT\rangle$ are found by equating the $\langle J\rangle$ expressions (1.24) with the vertical integrals of the $\overline{J}(z)$ expressions (1.23). The alternate expressions for $\overline{J}(z)$ found in this way are

$$\overline{J}(z) = \begin{cases} -\overline{T}'_B & = -\overline{T}'_T & = 1+\langle wT\rangle & \text{RB1} \\ -\overline{T}'_B+z & = -\overline{T}'_T-(1-z) & = \left(z-\frac{1}{2}\right)+\langle wT\rangle & \text{IH1.} \end{cases} \tag{1.25}$$

The constraints on $\overline{J}(z)$ and $\langle J\rangle$, expressed in terms of volume integrals, are summarized in the fifth and six rows of Tables 1.1 and 1.2.

In the IH1 configuration, there is yet another useful way to interpret $\langle wT\rangle$: in terms of the *fractions* of internally produced heat that flow outward across the top and bottom boundaries. Expressions for these fractions, which we call $\mathscr{F}_T$ and $\mathscr{F}_B$, follow from relations (1.21) and (1.25),

$$\mathscr{F}_T = -\overline{T}'_T = \tfrac{1}{2}+\langle wT\rangle \tag{1.26}$$

$$\mathscr{F}_B = \overline{T}'_B = \tfrac{1}{2}-\langle wT\rangle . \tag{1.27}$$

The top and bottom fractions are both 1/2 in the static state, but convective transport breaks this symmetry.

1.6.3 $\langle wT\rangle$ *and* $\delta\langle T\rangle$

Essential information about heat transport is captured by the volume integrals $\langle wT\rangle$ and $\delta\langle T\rangle$, where

$$\delta\langle T\rangle := \langle T-\overline{T}_T\rangle \tag{1.28}$$

is the mean fluid temperature, relative to that of the top boundary. The above definition is needed only for RB2 and IH2, where the top temperature is not fixed. In the other configurations, $\delta\langle T\rangle \equiv \langle T\rangle$ since we have set $T_T \equiv 0$. Neither $\langle wT\rangle$ nor $\delta\langle T\rangle$ is known a priori when the fluid is flowing. Instead, the quantities must be studied by physical and computational experiments, and they can sometimes be bounded analytically. All bounds and many experimental findings that we review in the following chapters can be stated in terms of $\langle wT\rangle$ or $\delta\langle T\rangle$. In the literature, however, results on $\langle wT\rangle$ are often stated differently but equivalently in terms of other quantities, including $\delta\overline{T}$, $\overline{T}'_B$, $\overline{T}'_T$, and the Nusselt number N defined in Sect. 1.6.4 below.

1.6.3.1   Uniform Bounds

The seventh and eighth rows of Tables 1.1 and 1.2 give uniform bounds on $\langle wT\rangle$ and $\delta\langle T\rangle$—that is, bounds that are independent of R and Pr. The bounds are derived in this chapter's appendix. The lower bounds on $\langle wT\rangle$ are tight in all six configurations. The upper bounds are thought to be tight (among uniform bounds), except in the IH2 case, where we suspect a uniform upper bound of 1/2. In the IH cases, the upper bounds on $\delta\langle T\rangle$ are tight, and the lower bounds are thought to be. In the RB cases, on the other hand, it is not clear whether any of the bounds on $\delta\langle T\rangle$ are tight.

The mean convective flux, $\langle wT\rangle$, saturates its lower bound of zero in each configuration only in the static state. Physically, this is because $\langle wT\rangle$ is proportional to the work exerted by buoyancy, and when motion persists this work must be positive to balance viscous dissipation. Mathematically, the positivity of $\langle wT\rangle$ in sustained convection follows from relation (2.37) in the next chapter. The upper bounds on $\langle wT\rangle$ correspond to limits in which convective transport is infinitely stronger than conductive transport. In the RB1 case, where $\delta\overline{T} = 1$, this limit is approached when $\langle wT\rangle$ grows without bound. In the four cases where the total heat flux, $\langle J\rangle$, is fixed, this limit is approached as $\langle wT\rangle \to \langle J\rangle$ (and $\delta\overline{T} \to 0$). The IH1 case is different in that $\delta\overline{T} = 0$, so $\langle wT\rangle$ is solely responsible for the mean vertical flux. However, if one thinks of the outward transport of heat across both boundaries, rather than upward transport, then the upper limit $\langle wT\rangle \to 1/2$ indeed means that convection fully takes over from conduction. The corresponding limits of the top and bottom flux fractions (1.26)–(1.27) are $\mathscr{F}_T \to 1$ and $\mathscr{F}_B \to 0$.

The mean temperature relative to that of the top boundary, $\delta\langle T\rangle$, is bounded above and below in all six cases, but the IH bounds differ in character from the RB1 bounds. The RB1 bounds given in Table 1.1 ensure that the mean temperature of the fluid lies between those of the top and bottom boundaries. The same may be true of $\delta\langle T\rangle$ in RB2 and RB3, but the bounds derived in this chapter's appendix are too crude to show it. In RB convection, the mean fluid temperature is exactly halfway between the boundary temperatures in the static state. The same is often true when the fluid is flowing, at least with symmetric boundary conditions, but it seems no rigorous statements have been proven that reflect this observation. In the IH cases, $\delta\langle T\rangle$ saturates the upper bounds of Table 1.2 only in the static states and is strictly smaller when the fluid is flowing. Its lower bound of zero, much like the uniform upper bounds on $\langle wT\rangle$, corresponds to convection being infinitely stronger than conduction. The R-dependent bounds of Sect. 2.3 show that $\delta\langle T\rangle$ could approach zero only as $R \to \infty$.

In IH convection, where R is proportional to the dimensional heating rate, H, it might seem counterintuitive that raising R tends to decrease $\delta\langle T\rangle$. However, the *dimensional* mean temperature, $\delta\langle T\rangle\Delta$, indeed rises with H. The dimensionless quantity $\delta\langle T\rangle$ falls as convection strengthens because it has essentially been normalized by its static value.

1.6.3.2 R-Dependent Bounds

Since many of the bounds on $\langle wT\rangle$ and $\delta\langle T\rangle$ given in Tables 1.1 and 1.2 are tight among uniform bounds, improving them requires finding bounds that depend explicitly on R or Pr. Some R-dependent bounds have been proven for the configurations we are considering, as summarized in Table 1.3. Bounds that vary with Pr have been proven recently for RB1 [12] but not yet for other cases, although some bounds have been proven for the infinite-Pr limit that are tighter than the corresponding uniform-in-Pr results, as discussed in Sect. 2.3.

In RB convection, the R-dependent bounds that have been proven are all upper bounds on $\langle wT\rangle$. They approach the uniform upper bounds as $R \to \infty$ but are tighter at all finite R. The uniform lower bounds of $0 \le \langle wT\rangle$ cannot be improved upon, if they are to hold for all solutions, since they are saturated by the static states. These states are unstable at large R, however, and $\langle wT\rangle$ typically grows with R in experiments and simulations. There might exist better lower bounds that hold only for attracting states, rather than all solutions, but we lack the mathematical machinery to prove them as yet.

In IH convection, the R-dependent bounds that have been proven are all lower bounds on $\delta\langle T\rangle$. They approach the uniform lower bounds of zero as $R \to \infty$ but are tighter at all finite R. No R-dependent upper bounds on $\langle wT\rangle$ have been proven in IH convection, but we argue in Sect. 1.6.3.4 that such proofs should be possible. On the other hand, the uniform upper bounds on $\delta\langle T\rangle$ and lower bounds on $\langle wT\rangle$ are saturated by the static states, so any efforts to improve them run into the same obstacle as efforts to improve the lower bounds on $\langle wT\rangle$ in RB convection.

Table 1.3 Proven bounds on $\langle wT\rangle$ and $\delta\langle T\rangle$ that hold at large R. The constant c differs between cases. Details and references are given in Sect. 2.3

	R-dependent bound on $\langle wT\rangle$	R-dependent bound on $\delta\langle T\rangle$
RB1	$\langle wT\rangle \le cR^{1/2}$	None
RB2, RB3	$\langle wT\rangle \le 1 - cR^{-1/3}$	None
IH1, IH2, IH3	None	$\delta\langle T\rangle \ge cR^{-1/3}$

1.6.3.3 Empirical Approximate Relations

In addition to the exact integral relations and bounds discussed above, experiments and simulations suggest some approximate relations for $\delta\langle T \rangle$ at large R. These relations are summarized in the ninth rows of Tables 1.1 and 1.2. Most can be expressed as relations between $\delta\langle T \rangle$ and $\langle wT \rangle$, at least approximately, but the underlying assumption in IH convection differs from that in RB convection.

In the RB cases, the underlying assumption is that the top and bottom boundary layers are roughly symmetric. This implies that the mean fluid temperature is about halfway between the top and bottom temperatures, as in the schematics of turbulent temperature profiles in Table 1.1. That is,

$$\delta\langle T \rangle \sim \tfrac{1}{2}\delta\overline{T} = \begin{cases} \tfrac{1}{2} & \text{RB1} \\ \tfrac{1}{2}\left(1 - \langle wT \rangle\right) & \text{RB2, RB3} \end{cases} \tag{1.29}$$

for large R. These relations are not expected to hold exactly when the top and bottom boundary conditions differ, but they could nonetheless be approached as $R \to \infty$. It might be possible to prove precise versions of the above statements, such as upper and lower bounds on $\delta\langle T \rangle$ that converge to $\delta\overline{T}$ as $R \to \infty$, but we are not aware of any such results.

In the IH cases, the underlying assumption is that mean temperature profile, $\overline{T}(z)$, at large R is roughly isothermal outside of thin thermal boundary layers, as in the schematics of turbulent temperature profiles in Table 1.2. Experimental support of this assumption is presented in Chap. 3. Approximate isothermally in the IH cases implies that

$$\delta\langle T \rangle \sim \begin{cases} \overline{T}_{\max} & \text{IH1} \\ \delta\overline{T} = \tfrac{1}{2} - \langle wT \rangle & \text{IH2, IH3} \end{cases} \tag{1.30}$$

for large R, where $\overline{T}_{\max}$ is the maximum of the mean temperature profile $\overline{T}(z)$. In IH1, the assumption of an isothermal interior does not give a relation between $\delta\langle T \rangle$ and $\langle wT \rangle$, nor is any simple relation suggested by experiments.

1.6.3.4 Conjectured Upper Bounds on $\langle wT \rangle$ in IH Convection

In IH2 and IH3, the empirical observation that $\delta\langle T \rangle \sim \delta\overline{T}$ at large R suggests that the two quantities might obey similar bounds. Since bounds of the form $\delta\langle T \rangle \geq cR^{-1/3}$ have been proven, it seems likely that bounds of the form $\delta\overline{T} \geq cR^{-1/3}$ could be proven also. The latter can be alternately stated as upper bounds on $\langle wT \rangle$:

Conjecture 1. In the IH2 and IH3 configurations, there exists a constant $c > 0$ such that for all Pr and sufficiently large R,

$$\langle wT \rangle \leq \tfrac{1}{2} - cR^{-1/3}.$$

In IH1, experiments suggest that the growth of $\langle wT \rangle$ with R is similarly bounded above (cf. Chap. 3). However, since no empirical relation between $\delta\langle T \rangle$ and $\langle wT \rangle$ is apparent, the proven lower bound on $\delta\langle T \rangle$ does not suggest a form for an upper bound on $\langle wT \rangle$. We speculate that the upper bound should approach the uniform bound of $1/2$ algebraically as $R \to \infty$, but we cannot anticipate the algebraic power:

Conjecture 2. In the IH1 configuration, there exist constants $c > 0$ and $\alpha > 0$ such that for all Pr and sufficiently large R,

$$\langle wT \rangle \leq \tfrac{1}{2} - cR^{-\alpha}.$$

1.6.4 Nusselt Numbers

The relative strengths of convective and conductive heat transport are often quantified using dimensionless Nusselt numbers. In RB convection, the typically used definitions of Nusselt numbers can all be expressed in terms of $\langle wT \rangle$. In IH convection, dimensionless quantities resembling the RB Nusselt numbers can be defined in various ways by invoking $\langle wT \rangle$, $\delta\langle T \rangle$, or $\overline{T}_{\max}$. Here we consider two ways of defining Nusselt-number-like quantities, N and $\tilde{N}$. The quantity N is determined by $\overline{T}_{\max}$ in IH1 and by $\langle wT \rangle$ in the other five cases, while the quantity $\tilde{N}$ is determined in the IH cases by $\delta\langle T \rangle$.

1.6.4.1 The Nusselt Number N

The definition of N that we choose is one that has helped reveal parallels between various RB configurations when used in concert with the quantity Ra defined in the next subsection [25, 36, 53, 54]. In every case other than IH1, our definition of N can be expressed as the ratio of mean total heat flux to mean conductive heat flux, where both quantities are averages over volume and time in the developed flow,

$$N = \frac{\langle J \rangle}{\langle J_{\text{cond}} \rangle} = \frac{\delta\overline{T} + \langle wT \rangle}{\delta\overline{T}} \qquad \text{RB1, RB2, RB3, IH2, IH3.} \qquad (1.31)$$

The above definition would fail for IH1 because its denominator would be zero. Applying the various constraints on $\delta\overline{T}$ and $\langle wT \rangle$ (cf. Tables 1.1 and 1.2) to expression (1.31) and adding an ad hoc definition for the IH1 case, we obtain

$$N := \begin{cases} 1 + \langle wT \rangle & \text{RB1} \\[2mm] \dfrac{1}{\delta\overline{T}} = \dfrac{1}{1 - \langle wT \rangle} & \text{RB2, RB3} \\[3mm] \dfrac{1}{8\overline{T}_{\max}} & \text{IH1} \\[3mm] \dfrac{1}{2\delta\overline{T}} = \dfrac{1}{1 - 2\langle wT \rangle} & \text{IH2, IH3.} \end{cases} \qquad (1.32)$$

The rationale for our definition of N in the IH1 case, the only case where heat flows outward across both boundaries, is that we are considering outward heat fluxes instead of upward fluxes. The mean total outward flux is unity since it must balance heat production. To determine the mean outward conduction, we imagine dividing the layer at the height z^* where the temperature profile $\overline{T}(z)$ assumes its maximum value of $\overline{T}_{\max}$. The upward conduction above z^* is proportional to $\overline{T}_{\max}$, as is the downward conduction below z^*. Thus, the ratio of total outward transport to conductive outward transport is inversely proportional to $\overline{T}_{\max}$. The $1/8$ factor makes N unity in the static state. Although the analogy between N in IH1 and in the other five cases is not perfect, the experiments reviewed in Chap. 3 reveal similarities in N between all cases. In the IH3 case, the quantity we call N has been considered under various names, perhaps first by Thirlby [49]. In the IH1 case, our definition has apparently not been used, but many authors have considered $\overline{T}_{\max}$, as well as the so-called top and bottom Nusselt numbers discussed in Sect. 3.2.2.

In all cases except IH1, our definitions (1.32) of N can be expressed in terms of $\langle wT \rangle$ alone. One might wonder whether N in the IH1 case would be better defined as inversely proportional to $1 - 2\langle wT \rangle$ instead of to $\overline{T}_{\max}$. This would be superficially identical to the IH2 and IH3 definitions of N, provided the latter are expressed in terms of $\langle wT \rangle$. However, the IH1 experiments discussed in Sect. 3.2 confirm that $\overline{T}_{\max}$ behaves very much like an inverse Nusselt number, while the quantity $1 - 2\langle wT \rangle$ does not. As described at the end of Sect. 1.6.2 above, $\langle wT \rangle$ in IH1 convection instead conveys the asymmetry between upward and downward heat fluxes.

1.6.4.2 The Nusselt Number $\tilde{N}$

Since $\delta\langle T \rangle$ is physically important in IH convection, it is natural to define a Nusselt-number-like quantity that is exactly related to $\delta\langle T \rangle$, rather than to $\langle wT \rangle$ or $\overline{T}_{\max}$. It works well to simply define $\tilde{N}$ as inversely proportional to $\delta\langle T \rangle$,

$$\tilde{N} := \begin{cases} \dfrac{1}{12\delta\langle T \rangle} & \text{IH1} \\[2mm] \dfrac{1}{3\delta\langle T \rangle} & \text{IH2, IH3.} \end{cases} \tag{1.33}$$

In the IH2 and IH3 cases, this definition could be interpreted as

$$\tilde{N} = \frac{\langle zJ \rangle}{\langle zJ_{\text{cond}} \rangle} \qquad \text{IH2, IH3,} \tag{1.34}$$

which is like the expression (1.31) for N with averages weighted proportionally to height. We do not define $\tilde{N}$ for RB convection, although the mean temperature in those cases merits attention also, as discussed in Sect. 1.6.3.

Table 1.4 Proven R-dependent bounds on N and $\tilde{N}$ that hold at large R. The constants c differ between cases. These are re-expressions of the bounds on $\langle wT \rangle$ and $\delta \langle T \rangle$ shown in Table 1.3

	R-dependent bound on $\langle wT \rangle$	R-dependent bound on $\delta \langle T \rangle$
RB1	$N \leq cR^{1/2}$	None
RB2, RB3	$N \leq cR^{1/3}$	None
IH1, IH2, IH3	None	$\tilde{N} \leq cR^{1/3}$

1.6.4.3 Basic Properties

In almost all cases it has been proven that $N \geq 1$ and $\tilde{N} \geq 1$, with equality holding only in the static states. These facts follow from the uniform bounds on $\langle wT \rangle$ and $\delta \langle T \rangle$ discussed in Sect. 1.6.3.1. It remains to be proven that $N \geq 1$ in the IH1 case, which would be true if $\overline{T}_{max}$ could not exceed its static value of 1/8. In turbulent convection, it is typically expected that $N \to \infty$ and $\tilde{N} \to \infty$ as $R \to \infty$. In the IH cases, this is tantamount to $\overline{T}_{max} \to 0$ or $\delta \langle T \rangle \to 0$. Such limiting behavior has not been proven but is supported by the experimental results described in Chap. 3.

Table 1.4 summarizes the R-dependent bounds that are known for N and $\tilde{N}$. These are simply restatements of the bounds on $\langle wT \rangle$ and $\delta \langle T \rangle$ given above in Table 1.3. In RB convection, the upper bounds on N are equivalent to upper bounds on $\langle wT \rangle$. In IH convection, the upper bounds on $\tilde{N}$ are equivalent to lower bounds on $\delta \langle T \rangle$. Upper bounds on N have not yet been proven for IH convection. In IH2 and IH3, bounds of the form $N \leq cR^{1/3}$ would follow from the upper bounds on $\langle wT \rangle$ that we have conjectured in Sect. 1.6.3.4. A bound of the same form for IH1 would require showing that $\overline{T}_{max}$ decays no faster than $R^{-1/3}$. The quantity $\overline{T}_{max}$ seems harder to access mathematically than volume averages like $\delta \langle T \rangle$ and $\langle wT \rangle$, which arise naturally in integral relations.

1.6.5 Diagnostic Rayleigh Numbers

The primary purpose of defining N and $\tilde{N}$ as we have is to identify similarities between the various configurations. The bounds in Table 1.4 suggest that we have almost succeeded, but the RB1 exponent is 1/2, while the others are 1/3. To bring the various cases completely into alignment, we must speak of the dependence of N and $\tilde{N}$ on *diagnostic* Rayleigh numbers, Ra and $\widetilde{Ra}$, instead of on the control Rayleigh number, R. These diagnostic parameters can be written simply as

$$Ra := \begin{cases} R & \text{RB1} \\ R/N & \text{RB2, RB3, IH1, IH2, IH3} \end{cases} \tag{1.35}$$

$$\widetilde{Ra} := R/\tilde{N} \quad \text{IH1, IH2, IH3.} \tag{1.36}$$

In terms of these variables, the RB bounds in Table 1.4 all take the form $N \leq cRa^{1/2}$, and the IH bounds take the form $\tilde{N} \leq c\widetilde{Ra}^{1/2}$. Moreover, the analogies brought out by considering N and Ra (or $\tilde{N}$ and $\widetilde{Ra}$) are not limited to bounds; similarities emerge also in experimental data [25, 53] and heuristic scaling arguments [19].

The different definitions of R, Ra, and $\widetilde{Ra}$ can be viewed as differences in the temperature scale used to define a Rayleigh number. The dimensional temperature scales Δ used to define R in Sect. 1.3 are characteristic of the static states, whereas Ra and $\widetilde{Ra}$ effectively replace Δ with temperature scales of the flowing fluid, Δ_{Ra} and $\Delta_{\widetilde{Ra}}$. In IH1, the temperature scale of Ra is the maximum horizontally averaged temperature in the flowing fluid. In the other five cases it is the mean temperature difference between the boundaries in the flowing fluid. That is,

$$\Delta_{Ra} = \begin{cases} \Delta & \text{RB1} \\ \delta\overline{T}\Delta & \text{RB2, RB3} \\ 8\overline{T}_{\max}\Delta & \text{IH1} \\ 2\delta\overline{T}\Delta & \text{IH2, IH3.} \end{cases} \tag{1.37}$$

Replacing Δ with Δ_{Ra} in the definition (1.10) of R leads to the above definition (1.35) of Ra. In the IH cases, the temperature scale of $\widetilde{Ra}$ is the volume-averaged temperature of the flowing fluid,

$$\Delta_{\widetilde{Ra}} = \begin{cases} 12\,\delta\langle T\rangle\Delta & \text{IH1} \\ 3\,\delta\langle T\rangle\Delta & \text{IH2, IH3.} \end{cases} \tag{1.38}$$

Replacing Δ with $\Delta_{\widetilde{Ra}}$ in the definition (1.10) of R leads to the above definition (1.36) of $\widetilde{Ra}$.

Appendix

In this appendix we prove various bounds on $\langle wT\rangle$ and $\delta\langle T\rangle$ that are uniform in the parameters, R and Pr. Most of these results are standard, but it is difficult to trace their origins, and we do not try.

Extremum Principles

In each configuration with $T = 0$ on a boundary, there holds a minimum principle giving pointwise, instantaneous lower bounds on $T(\mathbf{x}, t)$. For simplicity we assume that solutions to the Boussinesq equations exist and remain smooth. If $T(\mathbf{x}, t)$

ever achieves a local minimum on the interior, then at that point $\mathbf{u} \cdot \nabla T = 0$ and $\nabla^2 T \geq 0$, and so $\partial_t T \geq 0$. Thus, if the interior is initially warmer than the fixed boundary temperature of zero, it remains warmer for all time. Even if part of the interior is initially cooler than the boundary, it will be warmer than the boundary at large times. In the RB1 case, an analogous maximum principle holds relative to the warmer boundary, on which $T = 1$. For all $\mathbf{x}$ on the interior and sufficiently large t,

$$T(\mathbf{x},t) > 0 \qquad \text{RB1, RB3, IH1, IH3} \tag{1.39}$$

$$T(\mathbf{x},t) < 1 \qquad \text{RB1.} \tag{1.40}$$

In the RB2 and IH2 configurations, where fixed-flux conditions are imposed on both boundaries, neither maximum nor minimum principles hold pointwise.

Mean Convective Transport

Uniform bounds on the mean convective flux, $\langle wT \rangle$, in our RB and IH configurations are summarized in Tables 1.1 and 1.2. Many of these bounds follow from the power integrals (2.37)–(2.38), which are stated below for convenience.

$$\langle |\nabla \mathbf{u}|^2 \rangle = R \langle wT \rangle$$

$$\langle |\nabla T|^2 \rangle = \begin{cases} 1 + \langle wT \rangle & \text{RB1} \\ \delta \overline{T} = 1 - \langle wT \rangle & \text{RB2, RB3} \\ \delta \langle T \rangle & \text{IH1, IH2, IH3.} \end{cases}$$

In all six configurations, the $\mathbf{u}$ power integral implies $\langle wT \rangle \geq 0$. Since $\langle |\nabla \mathbf{u}|^2 \rangle > 0$ if convection persists, $\langle wT \rangle$ saturates its lower bound of zero if and only if the system approaches the static state as $t \to \infty$.

The uniform upper bounds on $\langle wT \rangle$ given in Tables 1.1 and 1.2 follow in most cases from lower bounds on $\delta \overline{T}$. In RB1 there is no upper bound on $\langle wT \rangle$. We get $\delta \overline{T} > 0$ from the T power integral in RB2 and RB3 and from the minimum principle in IH3. This lower bound on $\delta \overline{T}$ gives $\langle wT \rangle < 1$ in RB2 and RB3, where $\langle wT \rangle + \delta \overline{T} = 1$, and it gives $\langle wT \rangle < 1/2$ in IH1 and IH3, where $\langle wT \rangle + \delta \overline{T} = 1/2$. These upper bounds on $\langle wT \rangle$ are probably approached by certain solutions, including the turbulent attractors, as $R \to \infty$. If so, they are tight among uniform bounds. In IH1, the upper bound $\langle wT \rangle < 1/2$ follows from the minimum principle and relation (1.27).

It is likely that $\delta \overline{T} > 0$ in IH2 also, but we settle for the cruder estimate $\delta \overline{T} > -1/\sqrt{3}$, derived as follows.

$$\left|\delta\overline{T}\right| = \left|\langle\partial_z T\rangle\right|$$
$$\leq \langle|\partial_z T|\rangle$$
$$\leq \langle|\partial_z T|^2\rangle^{1/2}$$
$$\leq \langle|\nabla T|^2\rangle^{1/2}$$
$$\leq \delta\langle T\rangle^{1/2}$$
$$\leq \frac{1}{\sqrt{3}}.$$

The third line above follows from the Cauchy–Schwarz inequality, the fifth line follows from the T power integral, and the last line follows from the bound $\delta\langle T\rangle \leq 1/3$ that is proven below. The inequality is in fact strict since $\delta\langle T\rangle < 1/3$, except in the static state. Since $\langle wT\rangle + \delta\overline{T} = 1/2$ in IH2, the lower bound on $\delta\overline{T}$ gives $\langle wT\rangle < \frac{1}{2} + \frac{1}{\sqrt{3}}$.

In the IH1 configuration, we have also discussed $\overline{T}_{\max}$, the maximum value that $\overline{T}(z)$ attains. For this quantity, the lower bound $\overline{T}_{\max} > 0$ follows from the minimum principle, and the upper bound $\overline{T}_{\max} < 1/\sqrt{3}$ follows from a calculation very similar to the one in the previous paragraph. This upper bound is likely not tight; it might well be that $\overline{T}_{\max}$ never exceeds its static value of $1/8$.

Mean Temperature

Uniform bounds on the mean fluid temperature relative to that of the top boundary, $\delta\langle T\rangle$, are summarized in Tables 1.1 and 1.2. In the IH cases, the lower bounds $\delta\langle T\rangle > 0$ follow from the T power integral. The upper bounds are proven by integrating z^2 against the T equation (1.8), and using (1.26) in the IH1 case, to find

$$\delta\langle T\rangle = \begin{cases} \frac{1}{12} - \langle(z-\frac{1}{2})wT\rangle & \text{IH1} \\ \frac{1}{3} - \langle zwT\rangle & \text{IH2, IH3.} \end{cases} \tag{1.41}$$

Incompressibility gives $\overline{w} = 0$ and thus $\langle wT\rangle = \langle w\theta\rangle$ and $\langle zwT\rangle = \langle zw\theta\rangle$, where θ is the deviation of T from its static profile. Integrating the temperature fluctuation equation (2.3) against θ gives $\langle(z-\frac{1}{2})w\theta\rangle = \langle|\nabla\theta|^2\rangle \geq 0$ in IH1, and likewise for $\langle zw\theta\rangle$ in the other two cases. Therefore, $\delta\langle T\rangle \leq 1/12$ in IH1, and $\delta\langle T\rangle \leq 1/3$ in IH2 and IH3.

In the RB cases, none of the uniform bounds on $\delta\langle T\rangle$ are likely to be tight. The extremum principles give $0 < \delta\langle T\rangle < 1$ in RB1 and $0 < \delta\langle T\rangle$ in RB3. The upper bound for RB3 and the upper and lower bounds for RB2 follow from the inequality $|\delta\langle T\rangle| \leq 1/\sqrt{3}$ that is derived below.

$$|\delta\langle T\rangle| = |\langle z\partial_z T\rangle|$$
$$\leq \langle |z\partial_z T|\rangle$$
$$\leq \langle z^2\rangle^{1/2}\langle\partial_z T^2\rangle^{1/2}$$
$$\leq \tfrac{1}{\sqrt{3}}\langle|\nabla T|^2\rangle^{1/2}$$
$$\leq \tfrac{1}{\sqrt{3}}(1 - \langle wT\rangle)^{1/2}$$
$$\leq \tfrac{1}{\sqrt{3}}.$$

The first line of the derivation follows from integration by parts, the third line follows from the Cauchy–Schwarz inequality, the fifth line follows from the T power integral, and the last line follows from the bound $\langle wT\rangle \geq 0$ that is proven above.

References

1. Ahlers, G., Grossmann, S., Lohse, D.: Heat transfer and large scale dynamics in turbulent Rayleigh-Bénard convection. Rev. Mod. Phys. **81**(2), 503–537 (2009)
2. Ames, K.A., Straughan, B.: Penetrative convection in fluid layers with internal heat sources. Acta Mech. **85**, 137–148 (1990)
3. Asfia, F.J., Dhir, V.K.: An experimental study of natural convection in a volumetrically heated spherical pool bounded on top with a rigid wall. Nucl. Eng. Des. **163**(3), 333–348 (1996)
4. Aurnou, J., Andreadis, S., Zhu, L., Olson, P.: Experiments on convection in Earth's core tangent cylinder. Earth Planet. Sci. Lett. **212**, 119–134 (2003)
5. Avsec, D.: Tourbillons thermoconvectifs tans l'air. Application à la météorologie. Ph.D. thesis, Université de Paris (1939)
6. Calkins, M.A., Noir, J., Eldredge, J.D., Aurnou, J.M.: The effects of boundary topography on convection in Earth's core. Geophys. J. Int. **189**, 799–814 (2012)
7. Cardin, P., Olson, P.: Chaotic thermal convection in a rapidly rotating spherical shell: consequences for flow in the outer core. Phys. Earth Planet. Inter. **82**, 235–259 (1994)
8. Chandrasekhar, S.: Hydrodynamic and Hydromagnetic Stability. Dover, New York (1981)
9. Chapman, C.J., Childress, S., Proctor, M.R.E.: Long wavelength thermal convection between non-conducting boundaries. Earth Planet. Sci. Lett. **51**, 362–369 (1980)
10. Cheung, F.B., Chawla, T.C.: Complex heat transfer processes in heat-generating horizontal fluid layers. In: Annual Re view of Numerical Fluid Mechanics and Heat Transfer, vol. 1, pp. 403–448. Hemisphere, New York (1987)
11. Chillà, F., Schumacher, J.: New perspectives in turbulent Rayleigh-Bénard convection. Eur. Phys. J. E **35**(7), 1–25 (2012)
12. Choffrut, A., Nobili, C., Otto, F.: Upper bounds on Nusselt number at finite Prandtl number. arXiv:1412.4812v1 (2014)
13. Emanuel, K.A.: Atmospheric Convection. Oxford University Press, Oxford (1994)
14. Fearn, D.R., Loper, D.E.: Compositional convection and stratification of Earth's core. Nature **289**, 393–394 (1981)
15. Featherstone, N.A., Browning, M.K., Brun, A.S., Toomre, J.: Effects of fossil magnetic fields on convective core dynamos in A-type stars. Astrophys. J. **705**, 1000–1018 (2009)
16. Galdi, G.P., Straughan, B.: Exchange of stabilities, symmetry, and nonlinear stability. Arch. Ration. Mech. Anal. **89**(3), 211–228 (1985)

17. Gastine, T., Yadav, R.K., Morin, J., Reiners, A., Wicht, J.: From solar-like to antisolar differential rotation in cool stars. Mon. Not. R. Astron. Soc. Lett. **438**, 76–80 (2014)
18. Getling, A.V.: Rayleigh-Bénard Convection: Structures and Dynamics. World Scientific Publishing Co, Singapore (1998)
19. Goluskin, D., Spiegel, E.A.: Convection driven by internal heating. Phys. Lett. A **377**(1-2), 83–92 (2012)
20. Goluskin, D., Johnston, H., Flierl, G.R., Spiegel, E.A.: Convectively driven shear and decreased heat flux. J. Fluid Mech. **759**, 360–385 (2014)
21. Grötzbach, G., Wörner, M.: Direct numerical and large eddy simulations in nuclear applications. Int. J. Heat Fluid Flow **20**(3), 222–240 (1999)
22. Heimpel, M., Aurnou, J., Wicht, J.: Simulation of equatorial and high-latitude jets on Jupiter in a deep convection model. Nature **438**, 193–6 (2005)
23. Houseman, G.: The dependence of convection planform on mode of heating. Nature **332**, 346–349 (1988)
24. Hurle, D.T.J., Jakeman, E., Pike, E.R.: On the solution of the Bénard problem with boundaries of finite conductivity. Proc. R. Soc. A **296**(1447), 469–475 (1967)
25. Johnston, H., Doering, C.R.: Comparison of turbulent thermal convection between conditions of constant temperature and constant flux. Phys. Rev. Lett. **102**(6), 064501 (2009)
26. Jones, C.A.: A dynamo model of Jupiter's magnetic field. Icarus **241**, 148–159 (2014)
27. Joseph, D.D., Shir, C.C.: Subcritical convective instability: part 1. Fluid layers. J. Fluid Mech. **26**(4), 753–768 (1966)
28. Kaspi, Y., Flierl, G.R., Showman, A.P.: The deep wind structure of the giant planets: results from an anelastic general circulation model. Icarus **202**(2), 525–542 (2009)
29. Kippenhahn, R., Weigert, A.: Stellar Structure and Evolution. Springer, New York (1994)
30. Kolmychkov, V.V., Mazhorova, O.S., Shcheritsa, O.V.: Numerical study of convection near the stability threshold in a square box with internal heat generation. Phys. Lett. A **377**, 2111–2117 (2013)
31. Kulacki, F.A., Richards, D.E.: Natural convection in plane layers and cavities with volumetric energy sources. In: Natural Convection: Fundamentals and Applications, pp. 179–254. Hemisphere, New York (1985)
32. Lohse, D., Xia, K.Q.: Small-scale properties of turbulent Rayleigh-Bénard convection. Annu. Rev. Fluid Mech. **42**(1), 335–364 (2010)
33. Marshall, J., Schott, F.: Open-ocean convection: observations, theory, and models. Rev. Geophys. **37**, 1–64 (1999)
34. Nourgaliev, R.R., Dinh, T.N., Sehgal, B.R.: Effect of fluid Prandtl number on heat transfer characteristics in internally heated liquid pools with Rayleigh numbers up to 10^{12}. Nucl. Eng. Des. **169**, 165–184 (1997)
35. Oberbeck, A.: Ueber die wärmeleitung der flüssigkeiten bei berücksichtigung der strömungen infolge von temperaturdifferenzen. Ann. Phys. **243**(6), 271–292 (1879)
36. Otero, J., Wittenberg, R.W., Worthing, R.A., Doering, C.R.: Bounds on Rayleigh-Bénard convection with an imposed heat flux. J. Fluid Mech. **473**, 191–199 (2002)
37. Rajagopal, K.R., Ruzicka, M., Srinivasa, A.R.: On the Oberbeck-Boussinesq approximation. Math. Model. Methods Appl. Sci. **6**(8), 1157–1167 (1996)
38. Rayleigh, Lord: On convection currents in a horizontal layer of fluid, when the higher temperature is on the under side. Philos. Mag. **32**(192), 529–546 (1916)
39. Schubert, G., Turcotte, D.L., Olson, P.: Mantle Convection in the Earth and Planets. Cambridge University Press, Cambridge (2001)
40. Siggia, E.D.: High Rayleigh number convection. Annu. Rev. Fluid Mech. **26**, 137–168 (1994)
41. Soderlund, K.M., Schmidt, B.E., Wicht, J., Blankenship, D.D.: Ocean-driven heating of Europa's icy shell at low latitudes. Nat. Geosci. **7**(12), 16–19 (2014)
42. Sotin, C., Labrosse, S.: Three-dimensional thermal convection in an iso-viscous, infinite Prandtl number fluid heated from within and from below: applications to the transfer of heat through planetary mantles. Phys. Earth Planet. Inter. **112**, 171–190 (1999)

43. Sparrow, E.M., Goldstein, R.J., Jonsson, V.K.: Thermal instability in a horizontal fluid layer: effect of boundary conditions and non-linear temperature profile. J. Fluid Mech. **18**(04), 513–528 (1964)
44. Spiegel, E.A.: Thermal turbulence at very small Prandtl number. J. Geophys. Res. **67**(8), 3063–3070 (1962)
45. Spiegel, E.A.: Convection in stars I. Basic Boussinesq convection. Annu. Rev. Astron. Astrophys. **9**, 323–352 (1971)
46. Stern, M.E. (ed.): Ocean Circulation Physics. Academic Press, New York (1975)
47. Storey, B.D., Zaltzman, B., Rubinstein, I.: Bulk electroconvective instability at high Péclet numbers. Phys. Rev. E **76**(4), 041501 (2007)
48. Straughan, B.: Triply resonant penetrative convection. Proc. R. Soc. A **468**, 3804–3823 (2012)
49. Thirlby, R.: Convection in an internally heated layer. J. Fluid Mech. **44**(04), 673–693 (1970)
50. Thompson, P.A.: Compressible Fluid Dynamics. McGraw-Hill Inc, New York (1972)
51. van der Poel, E.P., Ostilla-Mónico, R., Verzicco, R., Lohse, D.: Effect of velocity boundary conditions on the heat transfer and flow topology in two-dimensional Rayleigh-Bénard convection. Phys. Rev. E **90**(1), 013017 (2014)
52. Vel'tishchev, N.F.: Convection in a horizontal fluid layer with a uniform internal heat source. Fluid Dyn. **39**(2), 189–197 (2004)
53. Verzicco, R., Sreenivasan, K.R.: A comparison of turbulent thermal convection between conditions of constant temperature and constant heat flux. J. Fluid Mech. **595**, 203–219 (2008)
54. Wittenberg, R.W.: Bounds on Rayleigh-Bénard convection with imperfectly conducting plates. J. Fluid Mech. **665**, 158–198 (2010)

Chapter 2
Stabilities and Bounds

Abstract Three configurations of internally heated convection and three configurations of Rayleigh–Bénard convection are analyzed mathematically. The standard
methods of determining the linear and energy stability thresholds of the static states
are explained. These analyses yield Rayleigh numbers above which the static states
are linearly unstable and Rayleigh numbers below which they are globally stable.
The resulting thresholds are reported with high precision for various boundary
conditions on the velocity. Exact analytical values, calculated by long-wavelength
asymptotics, are given for configurations with heat fluxes fixed at both boundaries.
It is then explained how the background method is used to prove lower bounds on
the mean fluid temperature in all three internally heated configurations, including
one for which no bound has been reported previously. In every configuration, these
bounds guarantee that the mean temperature of the fluid grows with the rate of
volumetric heating, H, no slower than $H^{2/3}$. The bounds are compared with upper
bounds on the Nusselt number in RB convection.

The preceding chapter has defined six convective configurations and described, for
each case, how various integral quantities characterize bulk heat transport. The
quantities of central importance include the mean vertical transport by convection,
$\langle wT \rangle$, in RB and IH convection and the mean temperature of the fluid relative to that
of the top boundary, $\delta \langle T \rangle$, in IH convection. We would like to predict the values
that these quantities assume for various Rayleigh numbers, Prandtl numbers, and
confining geometries. This is equivalent to predicting the parameter-dependence
of the Nusselt numbers we have defined; N is determined by $\overline{T}_{\max}$ in IH1 and
by $\langle wT \rangle$ in the other five cases, and $\tilde{N}$ is determined by $\delta \langle T \rangle$ in the three IH
cases. The task before us is very difficult in general and would require a greatly
improved understanding of fluid turbulence, so we are limited to partial results. The
present chapter presents facts about heat transport that can be determined purely by
mathematical analysis of the Boussinesq equations, while the next chapter addresses
simulations and laboratory experiments.

There are two main ways of studying heat transport analytically: examining
simple particular solutions that can be written down exactly or asymptotically, and
deriving bounds on integral quantities that apply to all solutions. The first method
yields much stronger results but is useful only at small Rayleigh numbers, where

© Springer International Publishing Switzerland 2016   27
D. Goluskin, *Internally Heated Convection and Rayleigh-Bénard Convection*,
SpringerBriefs in Applied Sciences and Technology,
DOI 10.1007/978-3-319-23941-5_2

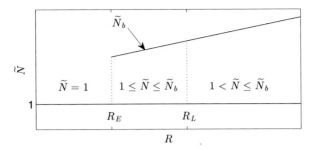

Fig. 2.1 Schematic of what this chapter's analytical results say about the dependence of $\tilde{N}$ on R in IH convection. The numerical values of R_E, R_L, and $\tilde{N}_b(R)$ vary between configurations. In RB convection, the analogous diagram for N lacks a middle region since $R_E = R_L$

the system either remains static or assumes a simple flow. Here we consider only the static states. Heat transport in static states is purely conductive and easy to understand, so the main task is to determine the parameters at which such states are stable. To this end, we can find a Rayleigh number, R_L, above which a static state is linearly unstable, and a Rayleigh number, R_E, below which we can prove that it is the unique globally stable state. These results work together with the second method of analysis—bounding N or $\tilde{N}$ above by functions $N_b(R)$ or $\tilde{N}_b(R)$—to constrain the dependence of Nusselt numbers on R. Sections 2.1–2.3 outline the calculations and values of R_L, R_E, and $\tilde{N}_b(R)$, respectively, for the various configurations.

The schematic of Fig. 2.1 shows how R_L, R_E, and $\tilde{N}_b(R)$ combine to give some knowledge of $\tilde{N}$ in IH convection. In the lowest-R regime, where $R < R_E$, we know that $\tilde{N} = 1$. This is because the system asymptotically approaches the static state, so its Nusselt number, which we have defined as an infinite-time limit, must be that of the static state. In the subcritical regime, where $R_E < R < R_L$, the static state is linearly stable, but sustained flow might also be possible, so all we can say is that $1 \leq \tilde{N} \leq \tilde{N}_b(R)$ in this regime. In the larger-R regime where $R_L < R$, the static state is linearly unstable, so any physically realizable state must have sustained flow and thus a Nusselt number strictly greater than unity. That is, $1 < \tilde{N} \leq \tilde{N}_b(R)$ for attracting states, although $\tilde{N} = 1$ remains possible if unstable states are included. A schematic like Fig. 2.1 for the other Nusselt number, N, in IH convection would lack the upper bound since R-dependent bounds on $\langle wT \rangle$ have not yet been proven (cf. Sect. 1.6.3.4).

Figure 2.1 represents a scenario where R_E is strictly smaller than R_L. In RB convection there is no subcritical regime since $R_E = R_L$. This allows for asymptotic solutions in the weakly supercritical regime, giving more precise expressions for Nusselt numbers there. In IH convection, on the other hand, subcritical convection is not ruled out because $R_E < R_L$. Stronger methods of analysis may be able to prove stability thresholds larger than R_E, but not necessarily as large as R_L. Subcritical convection is indeed possible in IH1 [42] and IH3 [3, 35, 43]. In IH2, the possibility of subcritical convection remains open.

Chapter 3
Biochips Fabrication and Surface Characterization

Abstract In this chapter, synthesized copolymer compositions are processed in order to fabricate coated biochips by using spin-coating technique. Developed biochips have subsequently been used in sandwich ELISA experiment via three different techniques: physical immobilization; covalent immobilization; and immobilization through amine-bearing spacers. In this chapter, principles behind each of the mentioned methods are described in a great detail. The major emphasis of the chapter is on the characterization of the developed biochips. In that path, morphology and topography of the coated platforms have thoroughly been investigated. Moreover, wettability of the polymer coated chips before and after treatment with different techniques were also investigated. Proposed platforms from both categories of treated and untreated, then, were analyzed by XPS analysis to study surface chemistry of the fabricated bioreceptors. This chapter has also discussed the correlation of obtained data from each characterization technique to another in a very comprehensive manner.

Keywords Physical immobilization · Covalent immobilization · Amine-bearing spacers · Morphology analysis · Topography analysis · Surface chemistry

3.1 Synthesis of Poly(MMA-co-MAA) Compositions

Four different compositions of poly(MMA-co-MAA) were synthesized via free-radical polymerization as shown in Table 2.1. Polymerization reaction was conducted in tetrahydrofuran (THF) as a solvent and by using azobisisobutyronitrile (AIBN) as an initiator. A detailed description of polymer synthesis can be found in our previous publications [1–4]. It is expected to have increasing concentration of –COOH groups as the concentration of MAA monomer in the polymerization reaction increases.

© The Author(s) 2016
S. Hosseini and F. Ibrahim, *Novel Polymeric Biochips for Enhanced Detection of Infectious Diseases*, SpringerBriefs in Forensic and Medical Bioinformatics, DOI 10.1007/978-981-10-0107-9_3

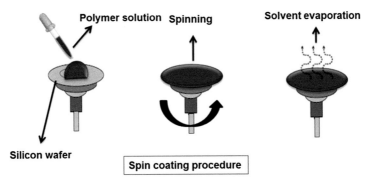

Fig. 3.1 Spin-coating technique for production of the polymer coated biochips [3]

3.2 Fabrication of Spin-Coated Biochips

Silicon wafers were used as substrates to produce a fine layer a polymer coated biochips by using spin-coating system (Fig. 3.1). Polymer solutions of different copolymer compositions (5 % solution) were prepared by dissolving polymers in THF.

A careful washing process is required prior to the spin-coating, which is explained in a great detail in our previous published works [2]. Variety of spin-coating parameters have been examined until the optimized method was determined [4]. One step coating with the duration of 55 s and the spinning rate of 3000 rpm was performed by Laurell, WS-650MZ-23NPP spin coater. The average thickness of the coated layer was ~10 μm. Since ELISA procedure includes many steps with long incubation times, the firmness of the developed coatings draws a great deal of importance. Polymer coated silicon biochips have shown significant stability as silicon wafer has shown a desirable adhesion property to hold the coated polymer layer on the surface even after long periods of incubation in aqueous medium.

3.3 Application of Biochips in ELISA and Surface Modification Techniques

Proposed coated surfaces in different poly(MMA-co-MAA) compositions (Fig. 3.2a) were cut into the dimension of 4 mm × 4 mm in order to fit perfectly at the bottom of the 96-well plate (Fig. 3.2b). As it was mentioned in Chap. 2, DENV from the family of NTDs has been chosen as the targeted virus for detection. Biomolecular immobilization and following DENV detection were performed via different methods: physical immobilization; covalent immobilization; and immobilization via amine-bearing spacers. For reminder, sandwich ELISA was chosen as

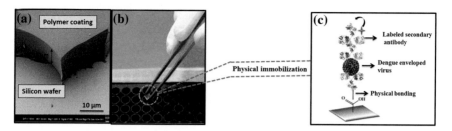

Fig. 3.2 Coated silicon wafer with the thickness of ~10 μm (**a**) were cut in the form of biochips to be placed into the ELISA kits (**b**); physical immobilization and subsequent DENV detection via direct attachment of the biomolecules to the surface (**c**) [3]

the protocol of choice for performing the assay. In this highly specific method, DENV is "sandwiched" between capture and primary antibodies. Primary antibody can further conjugate with labeled secondary antibody hence produces the detection signal.

3.3.1 Physical Immobilization

One of the most straight forward method for immobilization of biomolecules for subsequent biorecognition is physical immobilization [3]. This method relies on the direct attachment of the protein to the surface, which is commonly used in clinical practices such as ELISA (Fig. 3.2c). In this technique, immobilization mainly occurs through different interactions between targeted biomolecules and solid substrates [5, 6]. There are some key forces that can greatly influence the efficiency of the immobilization among which three important forces play vital roles in successful immobilization. Namely, ionic attraction (electrostatic interaction), hydrophobic interaction and hydrogen bonding are known to have a great impact on the immobilization efficiency [7]. Surface functional groups of the supporting substrates (such as –COOH in the present case) can interact with functional groups of the proteins (such as –NH$_2$) and result in protein attachment through ionic attraction [8, 9].

According to previous findings by Yoon et al. [7], the effect of ionic interaction on protein immobilization in comparison to other two forces can be considered insignificant. Hydrophobic nature of the supporting substrate can offer protein immobilization via hydrophobic interaction, while presence of surface –COOH groups (in the current case) can result in hydrogen bonding with the primary amines of the proteins as well [7]. Between these two major forces, hydrogen bonding has proven to influence protein immobilization in a stronger manner than hydrophobic interaction. Despite development of more complex immobilization techniques, clinical practice still relies on the direct attachment of the targeted biomolecules to the surface. This method is simple and cost effective as it avoids extra steps of modification and expensive chemicals used for treatments. Relative hydrophobicity of the bioreceptor surface hand in hand with the existence of desirable surface functional groups makes this method an effective strategy.

3.3.2 Covalent Immobilization

Covalent immobilization of antibody/antigen is one of the most commonly applied methods even in the routine laboratory diagnostics. In this method existing stable functionalities are generally transformed to the semi-stable highly reactive functional groups that can covalently bind to the proteins. Covalent immobilization can be performed via cross-linkers of different categories. A great number of research works have reported the application of zero-length cross linkers such as glutaraldehyde (GA), 1-ethyl-3-(3-dimethylaminopropyl) carbodiimide (EDC, appendix Table 3) and N-hydroxysuccinimide (NHS, appendix Table 4) in covalent immobilization of the proteins to the substrate [10, 11]. Functional conversion can occur quite differently from one cross-linking agent to another, thus making reproducibility difficult. For example, one of the most commonly used carbodiimide agents is the water-soluble EDC for aqueous crosslinking process. Carbodiimide such as EDC, can activate –COOH functional groups for direct conjugation to primary amines of the biomolecule. Available –COOH surface groups can be converted to unstable reactive O-acylisourea ester groups. This intermediate compound, by association of NHS can be converted to semi-stable NHS-ester groups (Fig. 3.3a), which is a highly reactive intermediate compound toward –NH$_2$ groups of peptide sequences (Lys) from proteins, thus resulting in covalent immobilization of the protein on the surface. However the versatility of generated functional groups in EDC/NHS chemistry makes repeatability of this technique difficult. Previous reports also indicate precautions and possible drawbacks of such approach as in some cases EDC/NHS reaction, under specific circumstances, can cause the formation of anhydride functional groups (instead of NHS-ester groups), which are unreactive towards proteins [11].

Early cross-linking inside the individual protein molecules might also happen when EDC/NHS treatment is aimed for activation of the surface. Such undesirable effects can cause the loss of protein activity that, in turn, result in poor detection signal and significant loss of sensitivity [12]. EDC and NHS as the representative cross-linking agents were used for covalent immobilization of biomolecules in this study. Biochips of different compositions were incubated in EDC/NHS solution (0.155 g of EDC and 0.115 g of NHS in 200 ml of PBS) for 1 h prior to capture antibody immobilization.

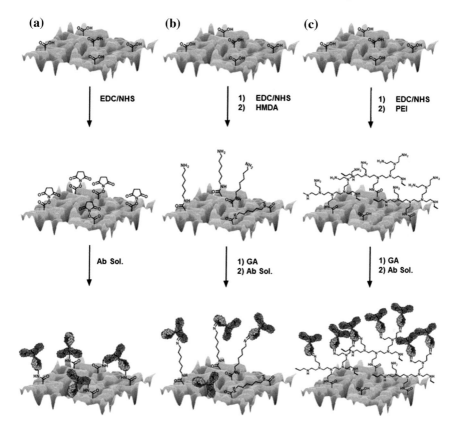

Fig. 3.3 Dengue antibody immobilization on the surfaces of the polymer coated biochips via: carbodiimide chemistry (**a**); and amine-bearing spacers such as HMDA (**b**) and PEI (**c**) [4]

3.3.3 Immobilization via Amine-Bearing Spacers

Larger cross-linkers are called "spacers molecules". Spacers are molecules with available functional groups for coupling with proteins. In particular, they can cause distance between proteins and the solid surface of the substrate, resulting in higher spatial freedom for binding proteins to the surface. Among all kinds of spacers, amine bearing molecules such as PEI (polyethylenimine, appendix Table 5), HMDA (hexamethylenediamine, appendix Table 6) and DAP (1,2-diaminopropane) are the most commonly used intermediate compounds [13]. Linear amine spacers such as HMDA are smaller in size and therefore offer less active functional groups in comparison to branched spacers such as PEI. For that reason, larger spacers normally result in higher immobilization rate due to the number of available active functionalities, which are offered for protein attachment. In the current study, HMDA (short aliphatic chain spacer) and PEI (branched spacer) were covalently attached to the surfaces of the biochips by the aim of carbodiimide chemistry. In the first step of modification, carbodiimide chemistry (as it was explained in Sect. 3.3.2) provides the reactive intermediate functionalities suitable for covalent attachment of the NHS ester groups to terminal $-NH_2$ functionalities of spacers (Fig. 3.3a, b). Surface amination was performed by incubation of the biochips in amine bearing-spacers solution for 1 h (0.1 % PEI in PBS or 0.1 % HMDA in PBS) followed by washing in PBS. In a further step, aminated surfaces react with glutaraldehyde (4 % in distilled water; 30 min), yielding aldehyde groups that could form imine linkages with primary amine groups of proteins [12, 14]. The amination procedure was conducted in ambient temperature [4]. Free $-NH_2$ groups of the spacer molecules were treated with GA solution to covalently attach dengue antibodies to the surface. GA is a well-known amine-reactive homobifunctional cross-linker, frequently used in biochemistry applications. Such an experimental strategy results in reactive functional groups for more robust protein binding with higher chance of reproducibility (Fig. 3.3b, c) [13, 14].

3.4   Morphology Analysis of the Biochips

Morphology of the developed biochips of different compositions was analyzed by SEM before and after treatment. A cross-section view of biochips (Fig. 3.4a) provides a perfect picture of the silicon substrate, coated layer of polymer and the gold coating on the top that was done to avoid surface charging during imaging. The average thickness of the coated layer was measured to be in the same range from one composition to another ($\sim 10 \ \mu m$).

Figure 3.5 depicts the frontal views of the coated biochips of 4 different compositions. Fine layers of polymer coatings can be observed on the surface of the silicon wafers. Such smooth and uniform coating layers, however, contain some cavities on the surface, which were formed as a result of solvent evaporation due to the low polymer concentration (Fig. 3.5).

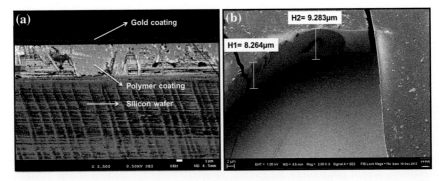

Fig. 3.4 Cross-section images of the coated biochips, PMAA coated chips were chosen as the representatives; **a** silicon wafer, the polymer coated layer, and the tin layer of gold coating on the top of the polymer layer; **b** thickness measurements for the polymer coatings

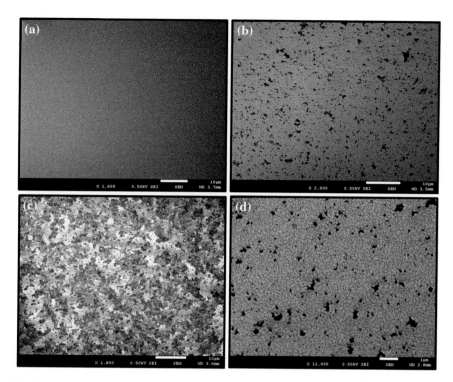

Fig. 3.5 Frontal SEM images of the polymer coated biochips by PMMA (**a**); comp.(9:1) (**b**); comp.(7:3) (**c**); and comp.(5:5) (**d**) [2]

To study the effect of surface treatment on the polymer coated surfaces, morphologies of the aminated biochips have also been investigated by SEM. Figure 3.6 presents frontal views of the PEI treated biochips, chosen as the representative modified samples [2].

The frontal view of PEI treated surfaces presents the smooth and uniform morphology for PMMA and comp.(9:1) coated biochips (Fig. 3.6a, b), while other two aminated surfaces contain irregular features (Fig. 3.6c, d). From the frontal view of the PEI treated comp.(5:5) obvious signs of delamination can be observed, which makes this composition somewhat disqualified for the assay. Furthermore, in detailed cross-section analysis of the mentioned samples, multilamination of the coated layers can be observed at the interface of coating and substrate, which, in turn, refers to the unstable presence of coated layer as a result of amination. Visual observations during the experiments have also shown that comp.(5:5) composition forms as a swollen gel-like material due to the abundance of MAA monomers, which are the hydrophilic segments of the copolymer [2, 15]. Similar delamination signs appeared on the HMDA treated biochips coated with comp.(5:5) [4].

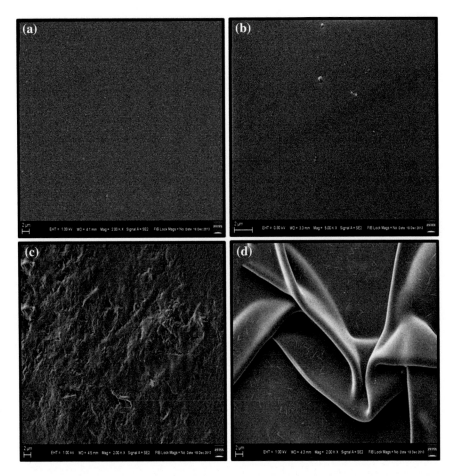

Fig. 3.6 Frontal SEM views of PEI treated polymer biochips coated by PMMA (**a**); comp.(9:1)
(**b**); comp.(7:3) (**c**); and comp.(5:5) (**d**) [4]

3.5 Topography Analysis of the Biochips

Detailed topography analysis of the biochips was performed by AFM and results are
presented in Fig. 3.7. AFM images reveal smooth surface features for PMMA coated
surfaces. Comp.(9:1) biochips have relatively rougher surface in comparison to
PMMA coated chips (Fig. 3.7a, b). Unlike relatively homogeneous surfaces of
PMMA and comp.(9:1), other two compositions depict heterogeneous rough features
on the surfaces of biochips coated with comp.(7:3) and comp.(5:5) (Fig. 3.7c, d).
Such surface topographies are greatly desired for the proposed application,
biomolecular interaction, as such structures provide higher chance of analyte-surface
interaction. Clear alteration in surface roughness of the coated chips is in direct
agreement with the changes in chemical structures of the coated surfaces. The higher

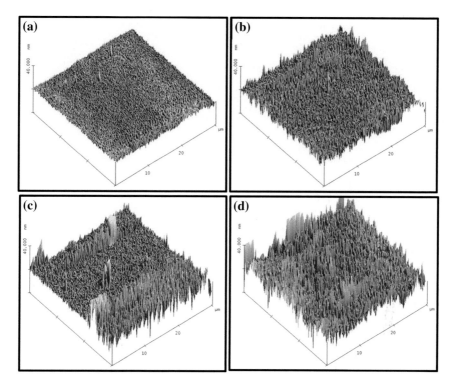

Fig. 3.7 Topography analysis of the surfaces of the biochips coated with: (**a**) PMMA; (**b**) comp. (9:1); (**c**) comp.(7:3); and (**d**) comp.(5:5) [2]

the concentration of the –COOH functional groups was tuned to produce the compositions, the higher surface roughness was recorded for the surfaces of the biochips. This is an obvious indication for the gradual increase of MAA ratio in polymerization reaction and subsequent presence of the –COOH groups in the compositions.

Further surface analysis has been dedicated to investigation of the aminated surfaces with AFM. In Fig. 3.8, representative images of biochips without treatment and after amination with HMDA and PEI are presented. Relatively large branched macromolecular structure of PEI in comparison to linear structure of HMDA has caused a clear alteration in the resultant topographies of the PEL treated biochips (Fig. 3.8c). Table 3.1 offers detailed information about mean roughness and root mean square roughness of the treated and untreated biochips also detected by AFM. Except PMMA coated biochips, aminated surfaces have shown significantly greater surfaces roughness compare to the untreated biochips (Table 3.1). A simple comparison between HMDA and PEI treated samples reveals that amination with PEI generally provides higher mean roughness and root mean roughness than HMDA.

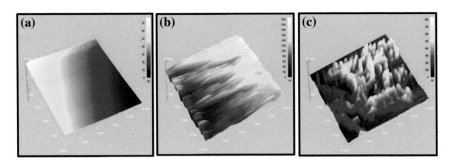

Fig. 3.8 Representative topography analysis of the biochips coated with comp.(9:1): (**a**) untreated; (**b**) HMDA treated; and (**c**) PEI treated comp.(7:3) [4]

Table 3.1 Detailed topography data determined by AFM [4]

Composition	Mean roughness (Ra, nm)	Root mean square roughness (Rq, nm)
PMMA	11.3	13.2
Comp.(9:1)	12.3	14.2
Comp.(7:3)	2.69	3.11
Comp.(5:5)	3.82	4.86
HMDA trearted coatings		
PMMA	10.7	13
Comp.(9:1)	24.6	31
Comp.(7:3)	43.2	51.7
Comp.(5:5)	90.2	112
PEI treated coatings		
PMMA	7.8	8.77
Comp.(9:1)	91.2	111
Comp.(7:3)	311	361
Comp.(5:5)	203	235

Greater surface roughness results in increased specific surface area available for biomolecular interaction. This, to a considerable extent, is expected to significantly influence the immobilization rate and subsequent biorecognition.

3.6   Wettability of the Biochips

Surface hydrophilicity of the biochips has been examined by water-in-air contact angle measurement and the results are presented in Fig. 3.9. The average contact angle measured for PMMA coated chips was found to be $\sim 78°$ that is in a good agreement with literature [16–18]. The gradual decrease in contact angle refers to

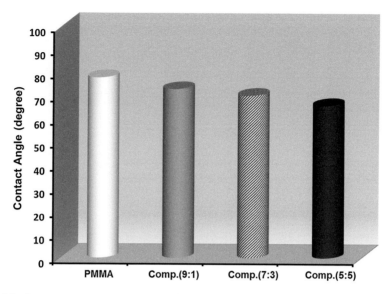

Fig. 3.9 Contact angles recorded for coated biochips [2]

the obvious chemical alterations of the examined surfaces. As the concentration of –COOH functional groups on the surfaces of the biochips increases, the respective contact angle evidently decreases, which is due to the hydrophilic nature of –COOH functionalities. Lowest contact angle, as expected, was measured for comp.(5:5) coated biochips ($\sim 66°$, Fig. 3.10d).

The contact angle analysis has also been performed for HMDA and PEI treated biochips and the results are shown in Fig. 3.11 [2]. In the case of aminated biochips, WCA was drastically decreased with respect to the increased surface concentration

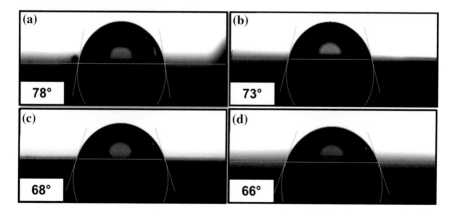

Fig. 3.10 Representative contact angle images recorded for coated surfaces with: (**a**) PMMA; (**b**) comp.(9:1); (**c**) comp.(7:3); and (**d**) comp.(5:5) [3]

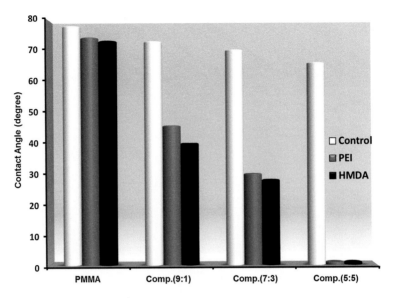

Fig. 3.11 Contact angles recorded for aminated biochips in comparison to controls (untreated samples) [4]

of –COOH groups, originated from controlled monomer contents. PMMA, inside the structure and on the surface, does not contain any –COOH functional groups available for binding to amine-bearing agents. Therefore, only a minor drop in WCA value ($\sim 5°$) was recorded for the aminated examined PMMA coatings in comparison to the rest. In contrast, comp.(5:5) coated platforms with the highest concentration of –COOH groups have shown super hydrophilic nature after treatment with both HMDA and PEI. Obtained data from WCA analysis shows clear agreement with results of AFM analysis [2, 4].

3.7 Surface Analysis of the Biochips by XPS

The chemical compositions at the surface of untreated and aminated polymer coated biochips were analyzed by XPS and the results are presented in Table 3.2. In the top row of the table, peak positions (eV) and chemical assignments are indicated ("-" refers to the concentration less than the detection limit: <0.2 %). The results obtained from the analysis of the untreated surfaces demonstrated a gradual decrease for the C1s peak (286.4 eV) upon increasing the concentration of MAA segments.

Such trend follows the logic as mentioned peak is attributed to carbon atoms of the –O–CH$_3$ groups that can be typically found in MMA building blocks. Therefore, it is expected that the intensity of this signal declines as compositions

Table 3.2 Surface analysis of the untreated and treated biochips [3, 4]

Peak	C1s				N1s		O1s
Binding energy (eV)	284.8	286.4	287.8	288.7	400.8	399.4	530
Peak assignment	–CH	–CO	–C=O	O–C=O	–N–C=O	–NH$_2$	
PMMA	42	16	–	12	–	–	26.5
Comp.(9:1)	44	15	–	13	–	–	26
Comp.(7:3)	45	13	–	12.5	–	–	27
Comp.(5:5)	48	10.5	–	11.5	–	–	26.5
HMDA treated coatings							
PMMA	42.5	17	–	11.5	–	0.8	25
Comp.(9:1)	42.5	15.5	1.7	10	–	1.7	25.5
Comp.(7:3)	42.5	17	1.8	4.5	–	4.6	24.5
Comp.(5:5)	37	14	3.1	2.6	–	3.1	32
PEI treated coatings							
PMMA	49.5	12	2.6	0.5	1	8.2	19.5
Comp.(9:1)	44	14.5	3	5	1.7	5.1	21
Comp.(7:3)	47.5	11.5	3.8	3.5	1.9	5.1	19
Comp.(5:5)	54.5	7.5	6.2	1	3.4	8.4	15.5

move from PMMA to comp.(5:5). A gradual decrease in MMA molar ratio is balanced with the increase in MAA concentration that, in turn, causes a natural increase in aliphatic carbon in comparison to the entire carbon atoms in the polymer chain (–CH, 284.8 eV). As it can be seen from Table 3.2, the intensity of this signal increases with the concentration of MAA monomers [3]. At the same time, from Table 3.2 it can be observed that the values for O–C=O peak (288.7 eV) and O1s peak (530 eV) have not revealed substantial alteration. Overall, in the case of untreated samples, XPS analysis confirms that –COOH functionalities exist on the outmost layer the surface and their concentration orderly increases with the concentration of MAA segments in polymer system [2, 3]. This is in line with our findings from AFM and WCA analysis discussed in Sects. 3.5 and 3.6.

In detailed surface analysis of aminated biochips, new peaks were identified on the surface of all treated platforms. The peak at 287.8 eV assigned to –C=O groups could be interpreted as amide bonds, which was detected on the surface of the biochips after carbodiimide coupling. As it can be seen in Table 3.1, there is no such group on the PMMA biochips, confirming the fact that formation of the amide groups took place only via conversion of the –COOH groups in EDC/NHS treatment. Other newly-detected peaks are N1s peaks. Aminated biochips are covered with a thin layer of, perhaps, few nanometers with aligned –NH$_2$ surface groups on the top (Table 3.2). In the case of HMDA treated samples, the concentration of –NH$_2$ groups increases as the MAA segments increases. For the composition with highest ratio of –COOH groups [comp.(5:5)], however, a decrease was detected in the surface concentration of –NH$_2$ after HMDA treatment. This specific platform has also shown an increased peak value for –C=O (287.8 eV). In the short linear structure of HMDA, there exist two –NH$_2$ groups on both sides of the structure. In abundance of the –COOH groups, both amine groups might react with the –COOH groups of the comp.(5:5) thus provide an undesirable effect, which limits the application of small spacer [4].

In the case of the PEI treated biochips, another new N1s peak at 400.8 eV appeared, which can be attributed to both imide (O=C–N–C=O) and amide (–N–C=O) functional groups [19, 20]. Detected imide groups can be produced via two different ways in the present case. They perhaps are the product of EDC/NHS reaction with surface –COOH groups that result in formation of NHS ester groups. As it was earlier mentioned in the text (Sect. 3.3.2), these functional groups are semi-stable and highly reactive towards –NH$_2$ groups of the spacers. Nonetheless, the chance that detected imide bonds are originated from carbodiimide reaction is minor as most likely by the time of the XPS analysis, almost all of the NHS ester groups have been hydrolyzed. Another possibility is that the EDC/NHS reaction with –COOH functionalities could also produce N-acylurea, which leads to the subsequent detection of imide bonds [11]. Proportional increase in atomic percentage of –NH$_2$ groups with C1s (–C=O) and N1s (–N–C=O) peaks indirectly confirms the amination efficiency, at which the ratio of amide bonds is relational to the ratio of available –NH$_2$ groups on the surface [19, 21].

Apart from chemical influence of the surface via abundance of exposed –NH$_2$ groups, hydrophilicity and surface roughness of the PEI treated biochips are expected to have a great impact on diffusion of the biomolecules to the surface hence providing higher binding efficiency with the aid of a amine-bearing spacers.

References

1. Hosseini S, Azari P, Farahmand E, Gan SN, Rothan HA, Yusof R, Koole LH, Djordjevic I, Ibrahim F (2015) Polymethacrylate coated electrospun PHB fibers: an exquisite outlook for fabrication of paper-based biosensors. Biosens Bioelectron 69:257–264. doi:10.1016/j.bios.2015.02.034
2. Hosseini S, Ibrahim F, Djordjevic I, Koole LH (2014) Polymethyl methacrylate-co-methacrylic acid coatings with controllable concentration of surface carboxyl groups: a novel approach in fabrication of polymeric platforms for potential bio-diagnostic devices. Appl Surf Sci 300:43–50. doi:10.1016/j.apsusc.2014.01.203
3. Hosseini S, Ibrahim F, Rothan HA, Yusof R, Marel Cvd, Djordjevic I, Koole LH (2015) Aging effect and antibody immobilization on –COOH exposed surfaces designed for dengue virus detection. Biochem Eng J 99:183–192. doi:10.1016/j.bej.2015.04.001
4. Hosseini S, Ibrahim F, Djordjevic I, Rothan HA, Yusof R, van der Mareld C, Koole LH (2014) Synthesis and processing of ELISA polymer substitute: the influence of surface chemistry and morphology on detection sensitivity. Appl Surf Sci 317:630–638. doi:10.1016/j.apsusc.2014.08.167
5. Rao S, Anderson K, Bachas L (1998) Oriented immobilization of proteins. Mikrochim Acta 128(3–4):127–143. doi:10.1007/BF01243043
6. YoungáJeong J, HyunáChung B (2008) Recent advances in immobilization methods of antibodies on solid supports. Analyst 133(6):697–701

7. Yoon J-Y, Park H-Y, Kim J-H, Kim W-S (1996) Adsorption of BSA on highly carboxylated microspheres—quantitative effects of surface functional groups and interaction forces. J Colloid Interface Sci 177(2):613–620

8. Alcon S, Talarmin A, Debruyne M, Falconar A, Deubel V, Flamand M (2002) Enzyme-linked immunosorbent assay specific to Dengue virus type 1 nonstructural protein NS1 reveals circulation of the antigen in the blood during the acute phase of disease in patients experiencing primary or secondary infections. J Clin Microbiol 40(2):376–381. doi:10.1128/jcm.40.02.376-381.2002

9. Xu H, Di B, Pan Y-x, Qiu L-w, Wang Y-d, Hao W, He L-j, Yuen K-y, Che X-y (2006) Serotype 1-specific monoclonal antibody-based antigen capture immunoassay for detection of circulating nonstructural protein NS1: implications for early diagnosis and serotyping of Dengue virus infections. J Clin Microbiol 44(8):2872–2878. doi:10.1128/jcm.00777-06

10. Wang C, Yan Q, Liu H-B, Zhou X-H, Xiao S-J (2011) Different EDC/NHS activation mechanisms between PAA and PMAA brushes and the following amidation reactions. Langmuir 27(19):12058–12068. doi:10.1021/la202267p

11. Sam S, Touahir L, Salvador Andresa J, Allongue P, Chazalviel JN, Gouget-Laemmel AC, Henry de Villeneuve C, Moraillon A, Ozanam F, Gabouze N, Djebbar S (2009) Semiquantitative study of the EDC/NHS activation of acid terminal groups at modified porous silicon surfaces. Langmuir 26(2):809–814. doi:10.1021/la902220a

12. Coad BR, Jasieniak M, Griesser SS, Griesser HJ (2013) Controlled covalent surface immobilisation of proteins and peptides using plasma methods. Surf Coat Technol 233:169–177

13. Bai Y, Koh CG, Boreman M, Juang Y-J, Tang IC, Lee LJ, Yang S-T (2006) Surface modification for enhancing antibody binding on polymer-based microfluidic device for enzyme-linked immunosorbent assay. Langmuir 22(22):9458–9467. doi:10.1021/la0611231

14. Wang Z-H, Jin G (2004) Covalent immobilization of proteins for the biosensor based on imaging ellipsometry. J Immunol Methods 285(2):237–243. doi:10.1016/j.jim.2003.12.002

15. Saunders BR, Crowther HM, Vincent B (1997) Poly[(methyl methacrylate)-co-(methacrylic acid)] microgel particles: swelling control using pH, cononsolvency, and osmotic deswelling. Macromolecules 30(3):482–487. doi:10.1021/ma961277f

16. Zhao B, Brittain WJ (1999) Synthesis of tethered polystyrene-block-poly(methyl methacrylate) monolayer on a silicate substrate by sequential carbocationic polymerization and atom transfer radical polymerization. J Am Chem Soc 121(14):3557–3558. doi:10.1021/ja984428y

17. Brown L, Koerner T, Horton JH, Oleschuk RD (2006) Fabrication and characterization of poly (methylmethacrylate) microfluidic devices bonded using surface modifications and solvents. Lab Chip 6(1):66–73. doi:10.1039/B512179E

18. Tennico YH, Koesdjojo MT, Kondo S, Mandrell DT, Remcho VT (2010) Surface modification-assisted bonding of polymer-based microfluidic devices. Sens Actuators B Chem 143(2):799–804. doi:10.1016/j.snb.2009.10.001

19. Kilian KA, Böcking T, Gaus K, Gal M, Gooding JJ (2007) Peptide-modified optical filters for detecting protease activity. ACS Nano 1(4):355–361. doi:10.1021/nn700141n

20. Xiao SJ, Textor M, Spencer ND, Wieland M, Keller B, Sigrist H (1997) Immobilization of the cell-adhesive peptide Arg–Gly–Asp–Cys (RGDC) on titanium surfaces by covalent chemical attachment. J Mater Sci Mater Med 8(12):867–872. doi:10.1023/A:1018501804943

21. Ghasemi M, Minier MJG, Tatoulian Ml, Chehimi MM, Arefi-Khonsari F (2011) Ammonia plasma treated polyethylene films for adsorption or covalent immobilization of trypsin: quantitative correlation between X-ray photoelectron spectroscopy data and enzyme activity. J Phys Chem B 115(34):10228–10238. doi:10.1021/jp204097a

Chapter 4
Application of Biochips in Dengue Virus Detection

Abstract This chapter is dedicated to the application of developed biochips in detection of dengue virus. Different techniques have been used for immobilization of dengue antibodies on the surface including: (i) physical attachment of the bio-molecules to the surface; (ii) biorecognition via covalent immobilization of the protein on the surface and; (iii) detection of dengue virus through different amine spacers. The performances of biochips made from different compositions have been compared in order to find the optimum surface among all the coated platforms. Conducted ELISA assay has been thoroughly evaluated in regards to essential parameters such as sensitivity, specificity, limit of detection and accuracy. Furthermore, involvement of the key forces such as ionic attraction, hydrophobic interaction and hydrogen bonding in analyte-surface interaction has been investigated with a major emphasis on surface properties of the developed biochips.

Keywords Dengue virus detection · Analyte-surface interaction · Hydrophobic interaction · Hydrogen bonding · Sensitivity · Specificity · Accuracy · Limit of detection

4.1 Physical Immobilization

4.1.1 Detection Range Study

The developed biochips coated with different copolymer compositions have been used in the sandwich assay (explained in Sect. 1.3). A broad range of virus concentration was chosen to study the performance of the respective biochips in the assay. Conventional ELISA has also been conducted on the exact same fashion to make the comparison complete (Fig. 4.1).

As the first observation it can be seen from Fig. 4.1 that the detection signal intensity gradually increased as the concentration of dengue virus decreased. This gradual increase in detection signal, however, was limited to the certain threshold at

© The Author(s) 2016
S. Hosseini and F. Ibrahim, *Novel Polymeric Biochips for Enhanced Detection of Infectious Diseases*, SpringerBriefs in Forensic and Medical Bioinformatics, DOI 10.1007/978-981-10-0107-9_4

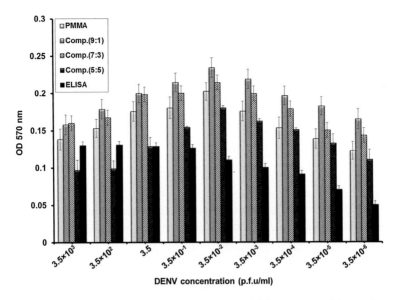

Fig. 4.1 Performance comparison between the biochips of different compositions and conventional ELISA via physical immobilization in a broad range of virus concentration [1]

which signal intensity switched the trend towards decreasing values [1]. This is most likely due to the large size of dengue antibody molecules and consequent loss of activity (denaturation) caused by steric repulsion. It is important to note that data in Fig. 4.1 were plotted after the cut-off values were first subtracted from the original data. Nevertheless, detection signals remained positive (regardless of the type of biochips) and consistently higher than conventional method even in the lowest concentrations of dengue virus.

4.1.2 Detection Performance Comparison

The data from representative virus concentration (DENV concentration $= 3.5 \times 10^{-2}$) was chosen and results are shown in Fig. 4.2. Negative controls have also been added to Fig. 4.2 for further clarification. Data in Fig. 4.2 indicates insignificant difference between performance of coated biochips with PMMA and comp.(7:3), while both compositions produced higher detection signal than conventional assay (Fig. 4.2, inset). The biochips of comp.(9:1) has shown better performance in DENV detection than rest of the examined samples. Although comp.(5:5) has also generated reasonable detection signals, its performance is not desirable considering relatively large errors produced by this particular composition. The reason for obtaining comparatively large negative controls for comp.(5:5) might be due to the gel-like characteristic of this compound as it was discussed is previous chapter (Sect. 3.7).

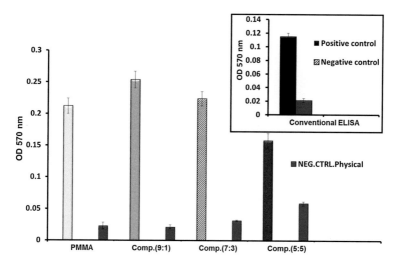

Fig. 4.2 Performance comparison of biochips from different compositions in DENV detection; *inset* signal intensity from conventional ELISA and its negative control

The swelling property of comp.(5:5) biochips could adsorb biomolecules into the polymer matrix in sandwich ELISA. Entrapped biomolecules would not be completely eliminated during the washing process thus partly remain inside the structure of the polymer coated layer of the chips. When the next biomolecule in the sequence is introduced to the assay, the chance of conjugation is significant which can subsequently lead to the generation of wrong signals. This fact, to a large extent, can result in large negative controls obtained for this particular composition in comparison to the rest [1–3].

Regardless of the detected negative controls, the resultant detection signal of comp.(5:5) biochips were found to be lowest among all the examined platforms. Therefore, it can be concluded that this composition is, perhaps, overly functionalized considering the fact that such phenomenon can cause an inefficient biomolecular interaction due to the steric hindrance. Overall, detection results clarify that, among all the coated platforms, comp.(9:1) biochips carry the optimal concentration of –COOH groups for most efficient protein immobilization. The previous results obtained from SEM analysis shows that comp.(9:1) has significant stability (Fig. 3.6) and thus this particular compositions is a qualified platform for the application in analytical assays.

4.1.3 Calibration Curves

The biochips with different compositions of poly(MMA-co-MAA) were examined in sandwich ELISA for DENV detection. Calibration curves were plotted (Fig. 4.3)

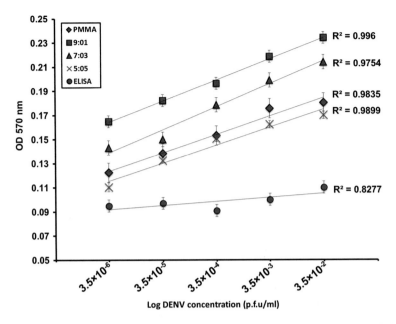

Fig. 4.3 *Calibration curves* obtained from sandwich ELISA conducted with biochips of different compositions as well as conventional ELISA [1]

by conducting the assay in DENV concentrations from 3.5×10^{-6} to 3.5×10^{-2} p.f. u/mL. Conventional ELISA has also been conducted in a parallel experiment in commercially available kits as control.

Figure 4.3 depicts higher level of precision resulted from polymer chips with the average square correlation coefficient of $R^2 \sim 0.98$ than conventional ELISA ($R^2 \sim 0.83$). Results also indicates that comp.(9:1) and comp.(7:3) with R^2 values of respectively 0.996 and 0.9754 have shown higher reliability than PMMA and comp.(5:5) biochips. In comparison to achieved results, conventional ELISA has performed in a poor manner showing the lowest level of precision with R^2 values of 0.8227.

4.1.4 Evaluation of the Assay

A careful evaluation of the assay was performed to determine essential parameter such as sensitivity, specificity and accuracy of the conducted assay (Table 4.1).

Total numbers of 128 replicates have been conducted including 96 positives and 32 negatives. Satisfactory results were obtained in the case of sensitivity of the developed platform as all the biochips have shown sensitivity of above 90 %. The greatest sensitivity was calculated for coated biochips with comp.(9:1) in comparison to the rest. The highest specificity, however, was found to be related to

Table 4.1 Sensitivity, specificity, accuracy and limit of detection (LoD) values calculated for the biochips of different compositions in DENV detection [1]

DENV status	PMMA		Comp. (9:1)		Comp. (7:3)		Comp. (5:5)		ELISA	
	+	−	+	−	+	−	+	−	+	−
Positive	88	3	94	2	91	1	89	2	69	12
Negative	8	29	2	30	5	31	7	30	27	20
Total	96	32	96	32	96	32	96	32	96	32
Sensitivity (%)	91.67		97.92		94.79		92.71		71.88	
Specificity (%)	90.62		93.75		96.88		93.75		62.50	
Accuracy (%)	91.4		96.87		95.31		92.96		69.5	
LoD (p.f.u × 10^3/mL)	16.7		1.22		0.963		41.21		500	

comp.(7:3) biochips, while rest of the compositions have also revealed a reasonable level of specificity. Moreover, comp.(9:1) occupied the highest level of accuracy as well. Supreme performance of comp.(9:1) and comp.(7:3) over other compositions is obvious from the detailed evaluation of the assay. Furthermore, LoD values were determined for coated platforms of all compositions and results are summarized in Table 4.1 as well. Proposed biochips successfully detected DENV in the satisfactory concentration level referred to the dengue patients at day 4–6 from onset of the DF [4]. Particularly, coated biochips with comp.(7:3) revealed the capability for DENV detection in lowest concentration levels (DENV concentration = 0.963×10^3 p.f.u/mL).

4.1.5   Chemistry Aspect

A general observation from the performance of different biochips has shown that the physical immobilization method resulted in consistently greater detection signals in comparison to conventional ELISA. The simplified method of direct immobilization was found to be largely reproducible and provides satisfactory level of reliability. Presence of polymer coating layers had an obvious impact on the detection efficiency in almost all the cases. Proposed biochips, either PMMA or different compositions of poly(MMA-co-MAA), are rich with oxygen atoms thus there is a dominant electronegativity on the surfaces of developed platforms. Such electronegativity can possibly involve ionic attraction and promote electrostatic interaction between –NH$_2$ groups of the proteins and the surface. In an extensive study performed by Yoon et al., major forces that can influence protein immobilization were classified. In this study, 3 major forces have been identified to have a great impact on analyte-surface interaction. Namely, ionic attraction, hydrophobic interaction and hydrogen bonding are these three forces, among which ionic attraction occupies less effective position in comparison to others [5]. This study also indicates that the effect of hydrogen bonding on biomolecular interaction is

Fig. 4.4 Major forces
involved in analyte-surface
interaction

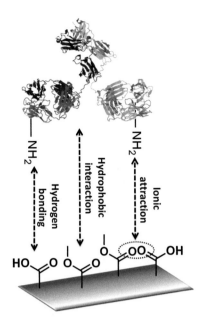

considerably greater than hydrophobic interaction [5]. Aligned with their findings, we have observed that PMMA coated biochips with the highest level of hydrophobicity performed in a relatively weaker manner than comp.(9:1) biochips, which are privileged with the presence of surface –COOH groups. Available surface –COOH groups promote hydrogen bonding with –NH₂ groups of the biomolecules (Fig. 4.4) [6]. This fact, to a great extent, emphasizes the crucial importance of functionally designed surfaces for their applications in biomolecular immobilization. Interface forces play fundamental and vital roles in analyte-surface interaction in general.

4.2   Covalent Immobilization

The immobilization of dengue antibody and detection of DENV have also been performed via covalent immobilization. In this method, by using carbodiimide chemistry approach, the biomolecules was covalently immobilized on the surface via EDC/NHS treatment. The detection signal intensities generated from different biochips are plotted and shown in Fig. 4.5. A simple comparison between performance of the developed platforms in physical (Fig. 4.2) and covalent (Fig. 4.5) immobilization confirms that there exist no major difference between two applied methods. It can be observed from the spreading (in special cases) that there is a higher chance of reproducibility for physical technique than covalent attachment. Physical attachment presents more reliable method, based on our observation, in

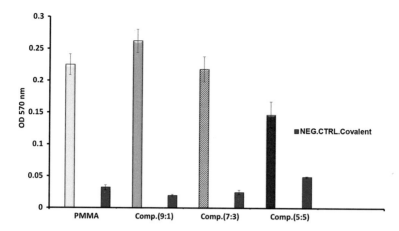

Fig. 4.5 Performance comparison of biochips from different compositions in DENV detection via covalent immobilization

comparison to covalently attachment of biomolecules to the surface, which normally involves expensive reagents and extra step of treatment in the assay.

Our result is in agreement with other researchers' findings [7, 8]. Sam et al. [7] claimed that in some cases, chemical immobilization of the biomolecules via EDC/NHS treatment is reported to be difficult to control and not always reproducible. Earlier reports have revealed that, under particular circumstances, EDC/NHS treatment might lead to the formation of anhydride functional groups, which are unreactive toward amine functional groups of the biomolecules [7, 8]. Frequent discussions have raised precautions and drawbacks of this treatment technique as such approach can result in an early cross-linking inside the individual biomolecules as well [9]. Such an undesirable phenomenon causes the loss of protein activity that subsequently results in generation of significantly lower detection signal [9]. Nonetheless, application of EDC/NHS treatment can still be beneficial in special cases.

4.3 Immobilization via Amine-Bearing Spacers

Surfaces of the coated biochips have been treated by chosen spacers prior to the assay. Linear diamine HMDA molecules were selected as the representative of small spacers. In contrast, PEI from the category of large branched spacers was chosen for the amination of the surfaces. As it was explained in pervious chapters, aminated biochips have been initially exposed to the glutaraldehyde and subsequently used for biomolecule immobilization [10, 11]. Detection performance of the different biochips under the influence of spacers is shown in Fig. 4.6. Statistical analysis of the detection performance is also presented in this Figure. In general, a

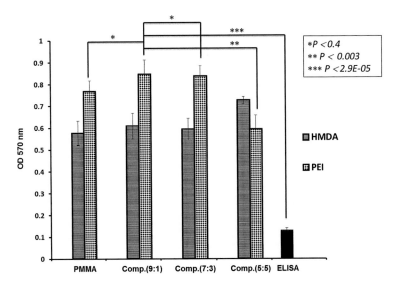

Fig. 4.6 Performance comparison of biochips from different compositions in DENV detection via amine-bearing spacers (HMDA and PEI). Conventional ELISA in detection of DENV is shown in *black*. Cut-off values are already excluded in plotting data [2]

substantial enhancement of the detection signal can be observed in the case of PEI treated surfaces in comparison to HMDA treated platforms. Branched structure of PEI molecules have affected the binding efficiency to a significant level as it provides abundance of $-NH_2$ groups for further analyte interaction [10, 12–14].

Although HMDA treated biochips performed in a relatively lower level of efficiency, detection signals are considerably higher than physical and covalent immobilization as well as the conventional assay (Fig. 4.6). The judgment between performances of two spacers is in a direct correlation with their sizes and the concentration of available $-NH_2$ groups for reaction. Considering the size of molecules, short spacers such as HMDA can only offer $-NH_2$ groups in a close proximity to the surface [10]. While, supreme performance of PEI molecules as spacer ($\sim$ 6–8 times greater signal intensity than conventional assay) can only be attributed to the presence of amine groups, which were generously distributed across the surface. In the case of comp.(5:5) biochips, however, the performance of the PEI spacer has considerably dropped. This is due to the overly functionalized surface of biochips that prevents efficient immobilization of the PEI spacer on the surface due to the steric repulsion. One may raise the question that why the same phenomenon did not occur when comp.(5:5) biochips were treated with HMDA. The answer can be found in the special structure of HMDA molecule as it offers only two units of $-NH_2$ groups per molecule thus avoiding steric hindrance (Fig. 3. 3). Such phenomenon proves that higher concentration of surface functional groups does not necessarily yield in more effective bio-functionalization of the platform. In contrast, only a careful control over the chemistry of the surface can provide the qualified platform for successful biomolecular interaction [3, 15].

References

1. Hosseini S, Ibrahim F, Rothan HA, Yusof R, van der Marel C, Djordjevic I, Koole LH (2015) Aging effect and antibody immobilization on –COOH exposed surfaces designed for dengue virus detection. Biochem Eng J 99:183–192
2. Hosseini S, Ibrahim F, Djordjevic I, Rothan HA, Yusof R, van der Mareld C, Koole LH (2014) Synthesis and processing of ELISA polymer substitute: the influence of surface chemistry and morphology on detection sensitivity. Appl Surf Sci 317:630–638
3. Hosseini S, Ibrahim F, Djordjevic I, Koole LH (2014) Polymethyl methacrylate-co-methacrylic acid coatings with controllable concentration of surface carboxyl groups: a novel approach in fabrication of polymeric platforms for potential bio-diagnostic devices. Appl Surf Sci 300:43–50
4. Thomas L, Najioullah F, Verlaeten O, Martial J, Brichler S, Kaidomar S, Moravie V, Cabié A, Césaire R (2010) Relationship between nonstructural protein 1 detection and plasma virus load in dengue patients. Am J Trop Med Hyg 83(3):696–699
5. Yoon J-Y, Park H-Y, Kim J-H, Kim W-S (1996) Adsorption of BSA on highly carboxylated microspheres—quantitative effects of surface functional groups and interaction forces. J Colloid Interface Sci 177(2):613–620
6. Ibrahim M, Nada A, Kamal DE (2005) Density functional theory and FTIR spectroscopic study of carboxyl group. Indian J Pure Appl Phys 44(12):911–917
7. Sam S, Touahir L, Salvador Andresa J, Allongue P, Chazalviel JN, Gouget-Laemmel AC, Henry de Villeneuve C, Moraillon A, Ozanam F, Gabouze N, Djebbar S (2009) Semiquantitative study of the EDC/NHS activation of acid terminal groups at modified porous silicon surfaces. Langmuir 26(2):809–814
8. Wang C, Yan Q, Liu H-B, Zhou X-H, Xiao S-J (2011) Different EDC/NHS activation mechanisms between PAA and PMAA brushes and the following amidation reactions. Langmuir 27(19):12058–12068
9. Situma C, Wang Y, Hupert M, Barany F, McCarley RL, Soper SA (2005) Fabrication of DNA microarrays onto poly(methyl methacrylate) with ultraviolet patterning and microfluidics for the detection of low-abundant point mutations. Anal Biochem 340(1):123–135
10. Bai Y, Koh CG, Boreman M, Juang Y-J, Tang IC, Lee LJ, Yang S-T (2006) Surface modification for enhancing antibody binding on polymer-based microfluidic device for enzyme-linked immunosorbent assay. Langmuir 22(22):9458–9467
11. Gunda NSKSM, Purwar Y, Shah SL, Kaur K, Mitra SK (2013) Micro-spot with integrated pillars (MSIP) for detection of dengue virus NS1. Biomed Microdevices 15:959–971
12. Coad BR, Jasieniak M, Griesser SS, Griesser HJ (2013) Controlled covalent surface immobilisation of proteins and peptides using plasma methods. Surf Coat Technol 233:169–177
13. Bucatariu F, Ghiorghita C-A, Simon F, Bellmann C, Dragan ES (2013) Poly(ethyleneimine) cross-linked multilayers deposited onto solid surfaces and enzyme immobilization as a function of the film properties. Appl Surf Sci 280:812–819
14. Gunda NSK, Singh M, Norman L, Kaur K, Mitra SK (2014) Optimization and characterization of biomolecule immobilization on silicon substrates using (3-aminopropyl)triethoxysilane (APTES) and glutaraldehyde linker. Appl Surf Sci 305:522–530
15. Hosseini S, Djordjevic I, Ibrahim F, Koole LH (2014) Recent advances in surface functionalization techniques on polymethacrylate materials for biosensor applications. Analyst

Summary

This book has described the synthesis of a novel copolymer system, poly (MMA-co-MAA), by different molar ratios of the monomers. High degree of control over concentration of generated functional groups has been achieved by variation of the monomers in polymerization reaction. Different compositions of the synthesized compolymer have been processed into biochips. The developed platforms in this study were employed for successful detection of dengue virus. Dengue fever is one of the most dangerous mosquito-borne viral infections mostly widespread in tropical and subtropical area. Detection of the analyte of interest (enveloped dengue virus) has been conducted through sandwich ELISA via three different techniques: (i) physical immobilization of the biomolecules on the surface by direct attachment of the dengue antibody on the biochips; (ii) dengue antibody immobilization by covalent attachment of the analyte on the surface performed by the aim of EDC/NHS treatment and; (iii) antibody immobilization and subsequent dengue virus detection through amine-bearing spacers. In all cases, novel polymeric platforms resulted in significantly improved detection signal and considerable enhancement of sensitivity, specificity and accuracy in comparison to the commonly applied assay, ELISA. A comprehensive characterization was performed on fabricated platforms and coated biochips have been analyzed by various techniques such as SEM, AFM, WCA and XPS. Performance of the biochips from different compositions of the copolymer were discussed and compared by major emphasis on the influence of surface micro-morphology and chemistry of the developed bioreceptors on the quality of biorecognition. Fabricated biochips in this study enhanced the detention signal between 6 to 8 times greater than conventional ELISA. Proposed methodology offers a ready-to-use functionalized surface with tailored surface properties that can promote analyte-surface interaction. Designed biochips not only carry the favourable functionalities but also hold the ideal concentration of such functional groups on the surface. More importantly, generated surface functionalities permanently exist on the surface as they are part of the chemical history of the material thus would not be affected by aging phenomena or deactivation of the surface due to the reorientation of the functional groups. For mentioned reasons, developed methodology is adequate for its application in biodiagnostics since it is privileged with long shelf-life and storability. Above all, fabricated biochips in this

© The Author(s) 2016

S. Hosseini and F. Ibrahim, *Novel Polymeric Biochips for Enhanced Detection of Infectious Diseases*, SpringerBriefs in Forensic and Medical Bioinformatics, DOI 10.1007/978-981-10-0107-9

study are capable of detecting other serotypes of dengue virus as well as variety of proteins. Polypeptides, in general, contain pendant carboxyl and amine groups that can be used for immobilization of such biomolecules to the surface. Therefore, presented approach can be used as a universal strategy for development of biosensors.

Future Investigations

This study has revealed great potentials for further improvement of the developed system. Since surface functional groups, in general, play a vital role in fundamental studies in biomaterials research, interaction of several cell types with surface gradients of functional groups can be an interesting field of research. It is known that cell-biomaterial interface has drawn a paramount importance in development of advanced biomaterials. Therefore, coated platforms with poly(MMA-co-MAA) could contribute in future direction of biomaterials. Variety of biomolecular entities can be immobilized on the surface in the similar fashion as dengue antibody. For instance, one of the potential analytes of interest can be tissue "growth factors" that have impact on cellular proliferation and differentiation after immobilization on solid platforms. Such approach can become one of the key solutions to ever-growing exploration of stem cells and tissue engineering applications in which substrate-cell interface plays the most significant role.

Furthermore, developed copolymer system in our study can be a potential candidate for fabrication of the new generation of analytical well plates replacing with the existing available kits. In specific, comp.(9:1) in its detailed characterizations, has offered stability along with optimal concentration of carboxyl groups for effective antibody immobilization. This transparent copolymer composition, which holds almost all of the characteristics of PMMA, could be a favourable material of choice for manufacturing new class of analytical kits with enhanced detection performance. Our analysis has proven that minor changes in the structure of materials from PMMA to poly(MMA-co-MAA) have greatly influenced the performance of the substrate.

While the developed biochips, in this study, enhanced the detection performance, yet proposed platforms have been produced by coating an expensive substrate such as silicon wafer. Silicon is a fragile material that limits its application. Therefore, such materials might not be the best choice for future industrialization of biochips. Furthermore, biochips are in the category of 2-dimensional materials, which have considerably smaller specific surface area available for biomolecular interactions in comparison to 3-dimensional platforms of the same materials. For that reason, fabrication of the spherical platforms made of the same copolymer system would offer significantly vast surface area that can subsequently provide higher chance of analyte-surface interaction.

Appendices

Methyl Methacrylate (MMA)

Methyl methacrylate is an organic compound (monomer) with known formula of $CH_2=C(CH_3)COOCH3$ (Appendix Table A.1). This colorless liquid can incorporate with the methyl ester groups of methacrylic acid (MAA) for production of poly (methyl methacrylate-co-methacrylic acid), poly(MMA-co-MAA), via free radical polymerization reaction.

Methacrylic Acid (MAA)

Methacrylic acid (MAA) is an organic monomer (Appendix Table A.2). This compound is also colorless and viscous liquid with an acrid unpleasant odor. It is

Table A.1 Additional information for methyl methacrylate

Methyl methacrylate	
IUPAC name: methyl 2-methylpropenoate	
Other names: MMA, 2-(methoxycarbonyl)-1-propene	
Identifiers	
CAS number	80-62-6
Properties	
Molecular formula	$C_5H_8O_2$
Molar mass	100.12 g mol^{-1}
Density	0.94 g/cm^3
Melting point	−48 °C, 225 K, −54 °F
Boiling point	101 °C, 374 K, 214 °F
Solubility in water	1.5 g/100 ml

© The Author(s) 2016
S. Hosseini and F. Ibrahim, *Novel Polymeric Biochips for Enhanced Detection of Infectious Diseases*, SpringerBriefs in Forensic and Medical Bioinformatics, DOI 10.1007/978-981-10-0107-9

Table A.2 Additional information for methacrylic acid

Methacrylic acid	
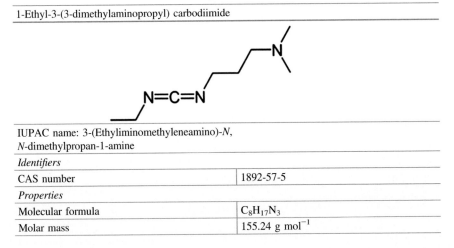	
IUPAC name: 2-methylpropenoic acid	
Other names: MAA, 2-methyl-2-propenoic acid	
Identifiers	
CAS number	79-41-4
Properties	
Molecular formula	$C_4H_6O_2$
Molar mass	86.06 g/mol
Density	1.015 g/cm^3
Melting point	14–15 °C
Boiling point	161 °C

soluble in warm water and miscible with most organic solvents. The presence of carboxyl groups in the structure of this monomer makes this compound desirable for surface engineering.

1-Ethyl-3-(3-dimethylaminopropyl) carbodiimide (EDC)

1-Ethyl-3-(3-dimethylaminopropyl) carbodiimide (EDC) is a water soluble carbodiimide agent (Appendix Table A.3). It is typically employed in the pH range of

Table A.3 Additional information for 1-Ethyl-3-(3-dimethylaminopropyl) carbodiimide

1-Ethyl-3-(3-dimethylaminopropyl) carbodiimide	
N=C=N structure	
IUPAC name: 3-(Ethyliminomethyleneamino)-*N*, *N*-dimethylpropan-1-amine	
Identifiers	
CAS number	1892-57-5
Properties	
Molecular formula	$C_8H_{17}N_3$
Molar mass	155.24 g mol^{-1}

4.0–6.0. General use of this compound is for activation of carboxyl groups for coupling with primary amines of proteins to yield amide bonds. EDC is normally used in combination with N-hydroxysuccinimide (NHS) or sulfo-NHS to increase coupling efficiency by creating a more reactive amine- product.

N-Hydroxysuccinimide (NHS)

N-Hydroxysuccinimide (NHS) is an organic compound with the known formula of $C_4H_5NO_3$ (Appendix Table A.4). NHS is slightly acidic, irritant to skin and eyes. NHS is commonly applied in biochemistry where it can be used as an activating reagent for converting carboxylic acids to intermediate amides groups for further coupling with proteins while a normal carboxylic acid would just form a salt with an amine.

Polyethylenimine (PEI)

Polyethylenimine (PEI) is a polymer compound in two forms of linear and branched structures (Appendix Table A.5). Linear polyethyleneimine contains all secondary amines. However, branched PEI contains primary, secondary and tertiary amino groups. Branched PEI can be polymerized by using ring opening reaction. Different degree of branching can be achieved by controlling the reaction parameters. PEI is known as a commonly used amine-bearing spacer for efficient protein immobilization.

Table A.4 Additional information for N-Hydroxysuccinimide

N-Hydroxysuccinimide	
IUPAC name: 1-Hydroxy-2,5-pyrrolidinedione	
Other names: 1-hydroxypyrrolidine-2,5-dione, HOSu	
Identifiers	
CAS number	6066-82-6
Properties	
Molecular formula	$C_4H_5NO_3$
Molar mass	115.09 g/mol
Appearance	Colorless solid
Melting point	95 °C, 368 K, 203 °F

Table A.5 Additional information for PEI

Polyethylenimine

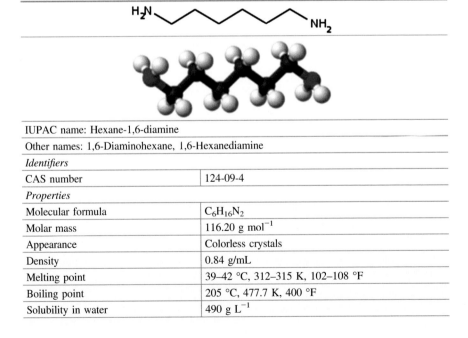

IUPAC name: Poly(iminoethylene)	
Other names: Polyaziridine, Poly[imino(1,2-ethanediyl)]	
Identifiers	
CAS number	9002-98-6
Properties	
Molecular formula	$(C_2H_5N)_n$, linear form
Molar mass	Variable

Hexamethylenediamine (HMDA)

Hexamethylenediamine (HMDA) is an organic compound with the formula of H_2N $(CH_2)_6NH_2$ (Appendix Table A.6). HMDA is a diamine molecule including a chain

Table A.6 Additional information for HMDA

Hexamethylenediamine

IUPAC name: Hexane-1,6-diamine	
Other names: 1,6-Diaminohexane, 1,6-Hexanediamine	
Identifiers	
CAS number	124-09-4
Properties	
Molecular formula	$C_6H_{16}N_2$
Molar mass	116.20 g mol^{-1}
Appearance	Colorless crystals
Density	0.84 g/mL
Melting point	39–42 °C, 312–315 K, 102–108 °F
Boiling point	205 °C, 477.7 K, 400 °F
Solubility in water	490 g L^{-1}

of hexamethylene terminated with amine functional groups. This colorless solid compound has a strong amine odor. HMDA has application in biomolecular interaction as an amine-bearing spacer.

Table 2.1 For no-slip boundary conditions: A Rayleigh number above which the static state is linearly unstable (R_L), a Rayleigh number below which the static state is globally stable (R_E), and upper bounds (N_b and $\tilde{N}_b$) on the Nusselt numbers N and $\tilde{N}$ that are valid for asymptotically large R. References for these values are given throughout the chapter. In the IH cases, R-dependent bounds on N have not been proven. In the RB cases, $\tilde{N}$ is not defined

	RB1	RB2	RB3	IH1	IH2	IH3
R_E	1707.76	720	1295.78	26,926.6	1429.86	2737.16
R_L	1707.76	720	1295.78	37,325.2	1440	2772.27
N_b	$0.027\,Ra^{1/2}$	$0.28\,Ra^{1/2}$	$0.28\,Ra^{1/2}$	None	None	None
$\tilde{N}_b$				$0.025\,\widetilde{Ra}^{1/2}$	$0.13\,\widetilde{Ra}^{1/2}$	$0.094\,\widetilde{Ra}^{1/2}$

For no-slip conditions on the velocity, Fig. 2.1 is made concrete in each configuration by Table 2.1, which gives values for R_L, R_E, $N_b(R)$, and $\tilde{N}_b(R)$. The bounds have been simplified by assuming that R is asymptotically large, and they are stated in terms of the diagnostic Rayleigh numbers, Ra and $\widetilde{Ra}$, that we have defined in Sect. 1.6.5. (Recall that Ra equals R in RB1 but equals R/N in the other five cases, and that $\widetilde{Ra} = R/\tilde{N}$ in IH convection.) The similarities between the various bounds are evident. The present chapter explains how the values in Table 2.1 are calculated and gives values for some other boundary conditions on the velocity.

The linear and nonlinear stability analyses that we apply to the static state can be applied to other particular solutions as well. Such analyses must be carried out asymptotically or numerically, however, since none of the finite-amplitude particular solutions can be expressed in closed form. The weakly nonlinear regime of IH convection has been theoretically examined in a few studies [34, 35, 40, 43]. Such analyses of particular solutions reveal much about bifurcations and pattern formation, but they do not yield robust information about heat transport. This is because the results often depend strongly on geometry, and also because each particular solution typically is stable over only a narrow range of parameters.

Sections 2.1 and 2.2 address each static state's linear and nonlinear stability, respectively. Section 2.3 outlines a proof of R-dependent lower bounds on the mean temperature for all three IH configurations. These bounds, which amount to upper bounds on $\tilde{N}$, are then compared with upper bounds on N in RB convection.

The results laid out in this chapter constitute most of what can be deduced mathematically about N and $\tilde{N}$. These results are rather meager in that they tell us neither the actual values assumed in the ranges $1 \le N \le N_b(R)$ and $1 \le \tilde{N} \le \tilde{N}_b(R)$, nor how the Prandtl number and geometry affect these values. Such questions must wait until the next chapter because substantial answers, so far, come only from simulations and laboratory experiments.

2.1   Linear Instability of Static States

In each RB and IH configuration, we can find a Rayleigh number, R_L, above which
the static state is linearly unstable. The Prandtl number of the fluid does not affect
this threshold. In most cases it has been proven that the linear instability is stationary,
meaning that the non-static states that bifurcate at the point of instability are steady,
rather than time-dependent. The method of calculating R_L is similar in every case
and is well known from the study of the canonical RB1 system. We outline this
methods here and give references for further details.

2.1.1   Linear Stability Eigenproblem

We want to study the stability of the static state, wherein $\mathbf{u} = \mathbf{0}$ and $T = T_{st}(z)$ for
the various $T_{st}(z)$ profiles given in expression (1.12). It is convenient to decompose
the temperature field into its static part and a fluctuation, θ,

$$T(\mathbf{x},t) = T_{st}(z) + \theta(\mathbf{x},t).$$

Since $\mathbf{u}$ and T evolve according to the Boussinesq equations (1.6)–(1.8), fluctuations
evolve according to

$$\nabla \cdot \mathbf{u} = 0 \tag{2.1}$$

$$\partial_t \mathbf{u} + \mathbf{u} \cdot \nabla \mathbf{u} = -\nabla p + Pr\nabla^2 \mathbf{u} + PrR\,\theta\hat{\mathbf{z}} \tag{2.2}$$

$$\partial_t \theta + \mathbf{u} \cdot \nabla \theta = \nabla^2 \theta - T'_{st}w, \tag{2.3}$$

where the prime denotes $\frac{d}{dz}$. The static state enters the fluctuation equations only
through its gradient, T'_{st}, reflecting the fact that Boussinesq dynamics are affected
only by relative temperature differences, not absolute temperatures. This gradient is
constant in RB convection but varies linearly in IH convection when the heating is
uniform:

$$T'_{st}(z) = \begin{cases} -1 & \text{RB1, RB2, RB3} \\ -z + \frac{1}{2} & \text{IH1} \\ -z & \text{IH2, IH3,} \end{cases} \tag{2.4}$$

where we recall that $0 \le z \le 1$. The boundary conditions on θ are the homogenous
analogs of the conditions on T:

$$\text{RB1, IH1:} \qquad \theta|_{z=0}, \quad \theta|_{z=1} = 0 \tag{2.5}$$

$$\text{RB2, IH2:} \quad \partial_z\theta|_{z=0}, \ \partial_z\theta|_{z=1} = 0 \tag{2.6}$$

$$\text{RB3, IH3:} \quad \partial_z\theta|_{z=0}, \ \theta|_{z=1} = 0. \tag{2.7}$$

The fluctuation dynamics of RB1 and IH1 are distinguished only by differing $T_{st}'(z)$. The same is true of RB2 and IH2, and of RB3 and IH3.

We study the stability of the zero solution of the fluctuation equations (2.1)–(2.3), which is equivalent to the stability of the static state. The nonlinear terms in the fluctuation equations can be neglected when finding linear stability thresholds. As is standard [4], we find a closed pair of equations governing the linear evolution of w and θ by taking $\hat{z} \cdot \nabla \times \nabla \times$ (2.2). Omitting time derivatives gives equations for the marginally stable states that are stationary, meaning they do not vary in time:

$$\nabla^4 w = -R\nabla_H^2\theta \tag{2.8}$$

$$\nabla^2\theta = T_{st}'w, \tag{2.9}$$

where $\nabla_H^2 := \partial_x^2 + \partial_y^2$ is the horizontal Laplacian operator. The validity of considering only stationary instabilities is discussed at the end of this subsection.

The Rayleigh number, R_L, at which the static state becomes linearly unstable is the smallest R for which there is a marginally stable state—that is, the smallest R for which Eqs. (2.8)–(2.9) have a nonzero solution. This is a (generalized) eigenproblem whose spectrum of eigenvalues is continuous and bounded below. Assuming there are no horizontal boundaries, we can Fourier transform the eigenproblem in x and y, decomposing it into an independent eigenproblem for each horizontal wavevector (k_x, k_y), where k_x and k_y are real. If the horizontal periods of a mode are L_x and L_y, then $k_x = 2\pi/L_x$ and $k_y = 2\pi/L_y$. The resulting decomposed eigenproblems take the form [4, 33]

$$\hat{w}^{(4)} - 2k^2\hat{w}'' + k^4\hat{w} = Rk^2\hat{\theta} \tag{2.10}$$

$$\hat{\theta}'' - k^2\hat{\theta} = T_{st}'\hat{w}, \tag{2.11}$$

where $\hat{w}(z)$ and $\hat{\theta}(z)$ are complex in general, and $k^2 := k_x^2 + k_y^2$. We call k the horizontal wavenumber.

The sixth-order linear system (2.10)–(2.11) requires six boundary conditions. The conditions (2.5)–(2.7) on θ apply also to $\hat{\theta}$. The first two $\hat{w}$ conditions are that $\hat{w} = 0$ at both boundaries, and the other two depend on whether each boundary is no-slip or free-slip:

$$\text{no-slip:} \ \hat{w}'(0), \ \hat{w}'(1) = 0 \tag{2.12}$$

$$\text{free-slip top:} \ \hat{w}'(0), \ \hat{w}''(1) = 0 \tag{2.13}$$

$$\text{free-slip bottom:} \ \hat{w}''(0), \ \hat{w}'(1) = 0 \tag{2.14}$$

$$\text{free-slip:} \ \hat{w}''(0), \ \hat{w}''(1) = 0. \tag{2.15}$$

Throughout this work, we give results for all four pairs of velocity conditions when possible. In some cases, analytical bounds and experimental results are available only for no-slip boundaries, which are the most natural in the laboratory. Condition (2.13) is experimentally realizable in a container with an open top. Condition (2.14) might appear unrealizable since it describes a container with an open bottom, but it also describes dynamically equivalent systems with an open top. For instance, IH convection with an open bottom, when viewed upside down, has the same dynamics as internally *cooled* convection with an open top.

For each k^2, Eqs. (2.10)–(2.11) and their boundary conditions form a linear eigenproblem in R with a *discrete* spectrum that is easier to compute than the continuous spectrum of (2.8)–(2.9). The R_L at which the static state loses stability is the smallest eigenvalue of (2.10)–(2.11), minimized over all admissible k^2. If all horizontal wavenumbers are possible,

$$R_L = \inf_{k^2 > 0} R^{(0)}(k), \qquad (2.16)$$

where

$$R^{(0)}(k) := \min \left\{ R \mid (2.10)\text{–}(2.11) \text{ has a nonzero solution} \right\}. \qquad (2.17)$$

The definition of R_L requires an infimum rather than a minimum because the infimum sometime occurs in the limit $k^2 \to 0$, in which case no minimum is achieved. Perturbations with $k = 0$ are not admissible since a horizontally uniform $\hat{w}$ would violate incompressibility. The value (or limit) of k at which R_L occurs is called the critical wavenumber of linear instability, k_L.

Because the eigenproblem (2.8)–(2.9) is derived assuming a stationary instability, the resulting R_L is the value at which a steady state bifurcates from the static one. In cases where it is proven that all marginally stable states are indeed stationary, $R > R_L$ is not only sufficient but necessary for instability of the static state. Stationarity has been proven for all RB cases [31]. In the IH3 case it follows for free-slip boundaries from an argument of Spiegel (see footnote 4 of [44]) and for no-slip boundaries from a theorem of Herron [16]. The latter method of proof may suffice to show stationarity in the remaining IH configurations. Until that is done, we can say in those cases only that $R > R_L$ is sufficient for instability.

2.1.2   Solutions of the Linear Stability Eigenproblem

In RB convection, where $T'_{st} = -1$, the eigenfunctions solving (2.10)–(2.11) are combinations of trigonometric and hyperbolic functions. The minimum eigenvalue at a given wavenumber, $R^{(0)}(k)$, must satisfy an expression involving the eigenfunctions. This expression can be solved for $R^{(0)}(k)$, analytically in a few cases and numerically in the others [4, 33]. In IH convection, where T'_{st} varies linearly

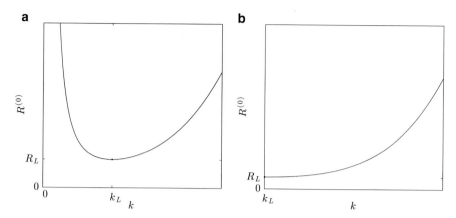

Fig. 2.2 Schematic diagrams of how the first marginally stable eigenvalue, $R^{(0)}$, depends on the horizontal wavenumber, k, in (**a**) the RB1, RB3, IH1, and IH3 cases, and in (**b**) the RB2 and IH2 cases, where heat fluxes are fixed at both boundaries

with z, the analogous approach would involve hypergeometric functions, so it is simpler to solve the eigenproblem (2.10)–(2.11) numerically. We have done this for all six configurations by the general method described in [41]: discretizing the differential operators using a spectral collocation method and computing the spectra of the resulting matrices. Our computed values of R_L agree with or add precision to the values in the literature. As explained shortly, the exact values of R_L in the RB2 and IH2 cases can be calculated also by asymptotic expansion.

The $R^{(0)}(k)$ curve can assume one of two qualitative shapes in our models, depending on the thermal boundary conditions. Figure 2.2a shows what the $R^{(0)}(k)$ curves look like in the four cases where the temperature is fixed at one or both boundaries: $R^{(0)}$ is minimized at a finite k and grows unboundedly both as $k \to 0$ and as $k \to \infty$. Figure 2.2b shows what the $R^{(0)}(k)$ curves look like when the temperature flux is fixed at both boundaries: $R^{(0)}$ approaches its infimum as $k \to 0$ and grows unboundedly as $k \to \infty$.

For all of our RB and IH configurations and all four pairs of velocity conditions (2.12)–(2.15), Table 2.2 gives the smallest Rayleigh number, R_L, at which the static state undergoes a stationary, horizontally periodic instability, along with the instability's horizontal wavenumber, k_L. Most of these values have been known for a long time. The RB1 case was analyzed first, with free-slip boundaries in Rayleigh's seminal analysis of 1916 [33] and then with other velocity conditions [20, 26, 31]. Later, values of R_L for some of our velocity conditions were reported for RB2 and RB3 [37], IH1 [11, 24, 37, 44, 45], IH3 [11, 24, 28, 34], and then IH2 [15, 19].

In most configurations, R_L is smallest when both boundaries are free-slip, largest when both boundaries are no-slip, and somewhere in between when one boundary is free-slip and the other is no-slip. The IH1 configuration provides a surprising exception: R_L is smallest when only the top is free-slip and largest when only the bottom is free-slip.

Table 2.2 Rayleigh number, R_L, above which each configuration's static state is linearly unstable, and the horizontal wavenumber, k_L, of the linear perturbation that is marginally stable at R_L. The RB2 and IH2 values are exact (cf. Sect. 2.1.3), while the other values are numerical approximations that are accurate to the precision shown

	R_L	k_L
RB1		
No-slip	1707.76	3.1163
Free-slip top	1100.65	2.6823
Free-slip bottom	1100.65	2.6823
Free-slip	657.511	2.2214
RB2		
No-slip	720	0
Free-slip top	320	0
Free-slip bottom	320	0
Free-slip	120	0
RB3		
No-slip	1295.78	2.5519
Free-slip top	816.744	2.2147
Free-slip bottom	668.998	2.0856
Free-slip	384.693	1.7576
IH1		
No-slip	37,325.2	3.9989
Free-slip top	16,669.8	3.0131
Free-slip bottom	37,949.4	4.0867
Free-slip	16,992.2	3.0277
IH2		
No-slip	1440	0
Free-slip top	576	0
Free-slip bottom	720	0
Free-slip	240	0
IH3		
No-slip	2772.27	2.6293
Free-slip top	1612.62	2.2611
Free-slip bottom	1650.55	2.1429
Free-slip	867.766	1.7897

In principle, knowing R_L is relevant to heat transport because $R > R_L$ suggests that sustained convection will occur. In confined geometries, however, not all k are admitted, and the flow can be only approximately periodic in the horizontal directions. These effects raise R_L by an amount particular to the confining geometry, and only when R exceeds this larger value is convection guaranteed.

The RB2 and IH2 configurations are special in that explicit expressions for R_L can be found analytically for any boundary conditions on the velocity. This is because the critical wavenumber is zero (cf. Table 2.2), so R_L can be calculated by long-wavelength asymptotics.

2.1.3 Long-Wavelength Asymptotics for RB2 and IH2

In the RB2 and IH2 cases, where heat fluxes are fixed at both boundaries, the infimum in the definition (2.16) of R_L occurs when $k \to 0$, so

$$R_L = \lim_{k \to 0} R^{(0)}(k). \tag{2.18}$$

It has apparently not been proven analytically that fixed-flux boundary conditions imply $k_L = 0$, so the finding must be verified on a case-by-case basis by numerically computing $R^{(0)}(k)$. This has been done for RB2 and IH2, so R_L can be found exactly by expanding the linear stability eigenproblem (2.10)–(2.11) in small k^2. (We cannot simply set $k = 0$ because the limit is singular.) The eigenmode with eigenvalue $R^{(0)}(k)$ has scaling $O(\hat{\theta}) = k^2 O(\hat{w})$ when $k \ll 1$ [6]. We thus let $\hat{w} = k^2 \hat{W}$ and seek solutions where $\hat{W}$ and $\hat{\theta}$ are both $O(1)$. The rescaled eigenproblem is

$$\hat{W}^{(4)} = R\hat{\theta} + 2k^2 \hat{W}'' - k^4 \hat{W} \tag{2.19}$$

$$\hat{\theta}'' = k^2 \left(\hat{\theta} + T_{st}' \hat{W} \right). \tag{2.20}$$

The no-flux conditions on $\hat{\theta}$ require the vertical integral of $\hat{\theta}''$ to vanish, and this furnishes a consistency condition on the right-hand side of (2.20):

$$\int_0^1 \left(\hat{\theta} + T_{st}' \hat{W} \right) dz = 0. \tag{2.21}$$

Equations (2.19)–(2.21) suffice to determine R_L, but it is possible to also retain the nonlinear terms in a long-wavelength expansion of the fluctuation equations. This has been carried out for RB2 [6], IH2 [19], and some other convective models with fixed boundary fluxes [5, 8]. The simpler linear calculation we describe here is contained in these nonlinear analyses.

Equations (2.19)–(2.20) are solved asymptotically by expanding in k^2:

$$\hat{W}(z) = W_0(z) + k^2 W_2(z) + \cdots \tag{2.22}$$

$$\hat{\theta}(z) = \theta_0(z) + k^2 \theta_2(z) + \cdots \tag{2.23}$$

$$R = R_L + k^2 R_2 + \cdots. \tag{2.24}$$

The R expansion anticipates that $R_0 = R_L$—in other words, that $k_L = 0$. We and others have confirmed this for RB2 and IH2 by computing the $R^{(0)}(k)$ curves numerically.

The procedure for asymptotically solving Eqs. (2.19)–(2.20) to any order k^{2n} is entirely systematic. Assuming all lower-order terms are known, the polynomial $\theta_{2n}(z)$ is found by integrating equation (2.20) at $O(k^{2n})$, then the polynomial $W_{2n}(z)$

is found by integrating equation (2.19) at $O(k^{2n})$, and finally R_{2n} is found from the consistency condition (2.21) at $O(k^{2n})$. Since $R_L = R_0$ here, we only need to carry out these three steps at leading order to find R_L. This has been done for RB2 in [6] and for IH2 in [15, 19]. The results of the three steps are that

$$
R_L = \frac{-1}{\int_0^1 T_{st}' P(z)dz} = \begin{cases} \dfrac{1}{\int_0^1 P(z)dz} & \text{RB2} \\[2ex] \dfrac{1}{\int_0^1 zP(z)dz} & \text{IH2,} \end{cases} \tag{2.25}
$$

where $P(z)$ is the unique fourth-order polynomial that has a leading coefficient of $1/24$ and satisfies the $\hat{w}$ boundary conditions. For our domain of $0 \le z \le 1$, these $P(z)$ are given in [15], for instance. The values of R_L for various velocity conditions appear in Table 2.2 above.

2.2   Energy Stability of Static States

Knowing the Rayleigh number, R_L, above which a static state is linearly unstable would be well complemented by knowing the critical Rayleigh number, R_c, below which it is the globally attracting state of the fully nonlinear dynamics. This is difficult in general, so we settle for finding a so-called *energy* Rayleigh number, R_E, that is a lower bound on R_c. Since linear instability implies nonlinear instability, we can anticipate that $R_E \le R_c \le R_L$. RB convection is special in that $R_E = R_c = R_L$ [21]. IH convection is more complicated in that $R_E < R_L$ for the largest known R_E, as depicted in Fig. 2.1. The values of R_E and R_L serve as upper and lower bounds on R_c, respectively, that hold uniformly for all Pr. The exact values of R_c can depend on Pr and are not yet known. The upper bounds on R_c can be tightened by finding particular subcritical solutions, as has been done for IH1 [42] and IH3 [35, 43], since any R at which subcritical convection persists must be larger than R_c. Proving a lower bound tighter than R_E is a more daunting challenge.

2.2.1   Lyapunov Stability and the Energy Method

The global stability of the static state is equivalent to the global stability of the zero solution to the fluctuation equations (2.1)–(2.3). The nonlinear terms in those equations that could be ignored in the linear stability analysis must now be included. The typical method of proving global stability, due to Lyapunov, requires finding a functional of the state variables that is nonnegative and whose evolution is nonpositive. That is, we must find a functional $\mathscr{L}[\mathbf{u}, \theta]$ such that

$$\mathscr{L}[\mathbf{u}, \theta] \geq 0 \tag{2.26}$$

$$\frac{d}{dt}\mathscr{L}[\mathbf{u}, \theta] \leq 0. \tag{2.27}$$

To show also that the static state attracts all initial conditions, it suffices for the above inequalities to be strict whenever $\mathbf{u}$ or θ is nonzero. In the convective systems we are studying, the best we can hope for is to find an $\mathscr{L}$ where (2.26) and (2.27) hold for R below some finite value, $R_{\mathscr{L}}$. That is,

$$R_{\mathscr{L}} := \sup\{R \mid \mathscr{L} \text{ satisfies (2.26)–(2.27)}\}. \tag{2.28}$$

The critical Rayleigh number R_c, is the largest R at which *any* Lyapunov functional exists,

$$R_c := \sup_{\mathscr{L}} R_{\mathscr{L}}. \tag{2.29}$$

There is no universally successful method for constructing Lyapunov functionals, let alone the optimal $\mathscr{L}$ that is valid for R up to R_c. It is even difficult to confirm that an optimal $\mathscr{L}$ is indeed optimal, except when $R_c = R_L$, as in RB convection. All we can do in general is make educated guesses for $\mathscr{L}$, determine the corresponding values of $R_{\mathscr{L}}$, and declare the largest $R_{\mathscr{L}}$ we can find to be a lower bound on R_c. In fluid dynamical systems like ours, even this guess-and-check procedure cannot be carried out for general $\mathscr{L}$ because it is too difficult to determine whether the second Lyapunov condition (2.27) holds. In most nonlinear analyses of fluid stability, this trouble is avoided by considering only a particular subset of possible Lyapunov functionals for which it is tractable to check the second Lyapunov condition. This approach is called the *energy method*.

The energy method in fluid mechanics [22, 36, 39] is a special case of Lyapunov's method. In one definition of the energy method that is neither the narrowest nor the broadest definition possible, the Lyapunov functional, which is called the energy, has two special features:

1. The energy is quadratic in the state variables.
2. The energy is conserved by the nonlinear terms of the fluctuation equations (2.2)–(2.3), meaning that these terms do not contribute to the expression for the time-evolution of the energy.

The energy method is so named because quadratic quantities are often proportional to physical energies. Here we follow Joseph [21] in considering energies of the form

$$E_\gamma[\mathbf{u}, \theta](t) := \frac{1}{2} \int \left(\frac{1}{PrR}|\mathbf{u}|^2 + \gamma\theta^2\right) d\mathbf{x}, \tag{2.30}$$

where $\int$ denotes an instantaneous volume average. The constant $\gamma > 0$ is called a *coupling parameter*, and each positive value defines an energy that is a valid Lyapunov functional for R up to some R_{E_γ}. This value of R_{E_γ} is maximized by some optimal choice of γ, where it achieves the critical Rayleigh number of energy stability, R_E:

$$R_E := \max_{\gamma > 0} \sup \left\{ R \mid E_\gamma \text{ satisfies } (2.26)-(2.27) \right\}. \qquad (2.31)$$

The value of R_E is the best lower bound on R_c that we find by the energy method, though it is likely still smaller than R_c. Deriving a better lower bound on R_c would require going beyond the energy method to search over a larger class of Lyapunov functionals, and this presents technical challenges. Progress beyond the energy method has been made for a few shear flow models [7, 23] but not yet for a convective system.

2.2.2   Energy Stability Eigenproblem

The functional E_γ suffices to show that the static state is globally stable whenever it satisfies conditions (2.26)–(2.27). The first condition holds whenever all the parameters are positive, so it remains only to determine the parameters for which $\frac{d}{dt} E_\gamma \leq 0$. Adding the volume averages of $\frac{1}{Pr R} \mathbf{u} \cdot (2.2)$ and $\gamma \theta \times (2.3)$ and then integrating by parts gives

$$\tfrac{d}{dt} E_\gamma = - \int \left[\tfrac{1}{R} |\nabla \mathbf{u}|^2 + \gamma |\nabla \theta|^2 - \left(1 - \gamma T'_{st} \right) w\theta \right] d\mathbf{x}. \qquad (2.32)$$

The static state is globally attracting if the right-hand side of (2.32) is negative definite. Like the linear stability threshold, the satisfaction of this condition depends on R but not on Pr. Only static states have this feature; other solutions and their stabilities depend also on Pr.

   The calculus of variations yields a necessary and sufficient condition for the righthand side of (2.32) to be negative definite. In particular, E_γ is a Lyapunov functional when R is smaller than all eigenvalues, R, of the (generalized) eigenproblem [1, 38, 39]

$$\hat{w}^{(4)} - 2k^2 \hat{w}'' + k^4 \hat{w} = \tfrac{1}{2} R k^2 \left(1 - \gamma T'_{st} \right) \hat{\theta} \qquad (2.33)$$

$$\gamma \left(\hat{\theta}'' - k^2 \hat{\theta} \right) = -\tfrac{1}{2} \left(1 - \gamma T'_{st} \right) \hat{w}. \qquad (2.34)$$

The boundary conditions are the same as in the linear stability eigenproblem of Sect. 2.1.2, and again $\hat{w}(z)$ and $\hat{\theta}(z)$ can be complex, and k is the horizontal wavenumber. Expression (2.31) for R_E can thus be restated as

$$R_E = \max_{\gamma > 0} \inf_{k^2 > 0} \min \left\{ R \mid (2.33)-(2.34) \text{ has a nonzero solution} \right\}. \qquad (2.35)$$

It is a special feature of the energy method, and not of Lyapunov's method in general, that the nonlinear stability analysis can be reduced to the solution of a linear eigenproblem, much like the linear stability analysis.

2.2.3   Solutions of the Energy Stability Eigenproblem

In RB convection, where $T'_{st} = -1$, there is no need to solve the energy stability
eigenproblem (2.33)–(2.34) because it is identical to the linear stability eigenproblem (2.10)–(2.11), so long as the energy is defined with $\gamma = 1$. This energy
is thus a valid Lyapunov functional for all R up to R_L. (The agreement of the
two eigenproblems reflects the self-adjointness of the linear stability operator; see
[14, 39].) We expect $\gamma = 1$ to be the optimal coupling parameter since R_E should
not exceed R_L, and indeed this can be shown directly [21]. These observations
justify our earlier assertion that $R_E = R_c = R_L$ in RB convection, making subcritical
instability impossible.

In IH convection, R_E must be calculated by performing the double optimization
of expression (2.35), which requires solving the eigenproblem (2.33)–(2.34). In all
IH cases the strict inequality $R_E < R_L$ holds. Table 2.3 gives values of R_E for the
various IH configurations, along with the percent differences between R_L and R_E,
and the arguments, γ^* and k_E, that yield the maxima and infima in expression (2.35).

The relative magnitudes of the gaps between R_E and R_L are on the order of 1% in
IH2 and IH3, where the bottom is insulating, but are much larger in IH1, where heat

Table 2.3 Rayleigh number (R_E) below which the energy method
proves that each IH configuration's static state is globally attracting, the percentage of R_L by which R_E falls short of R_L, the
optimal coupling parameter (γ^*) used to define the energy that is
a valid Lyapunov functional for all $R < R_E$, and the horizontal
wavenumber (k_E) at which the infimum in (2.35) occurs for the
optimal energy. The IH2 values are numerical approximations to
the exact analytical expressions (2.36). The IH1 and IH3 values are
computed numerically and are accurate to the precision shown

	R_E	% below R_L	γ^*	k_E
IH1				
No-slip	26 926.6	27.9	8.8831	3.6174
Free-slip top	12 620.2	24.3	7.9626	2.9014
Free-slip bottom	24 722.8	34.9	9.1975	3.3664
Free-slip	10 618.1	37.5	8.8516	2.5498
IH2				
No-slip	1429.86	0.704	1.9720	0
Free-slip top	573.391	0.453	1.7838	0
Free-slip bottom	714.929	0.704	2.2185	0
Free-slip	239.055	0.394	1.9843	0
IH3				
No-slip	2737.16	1.27	2.0678	2.6355
Free-slip top	1594.42	1.13	1.9185	2.2661
Free-slip bottom	1624.26	1.59	2.3702	2.1512
Free-slip	855.674	1.39	2.1821	1.7958

escapes across both boundaries. We cannot say whether the larger gaps in IH1 are necessitated by subcritical solutions or are only mathematical artifacts of the optimal energies being poor approximations of the truly optimal Lyapunov functionals. The most energy-unstable wavenumber, k_E, is fairly close to k_L in IH1 and IH3, and $k_E = k_L = 0$ in IH2. The optimal coupling parameters, γ^*, are all significantly larger than unity, which is their optimal value in the RB cases.

For the IH1 and IH3 cases, we numerically computed the values of R_E, γ^*, and k_E that appear in Table 2.3 using the same spectral collocation method that we used to compute R_L. A similar energy stability analysis was carried out by Kulacki and Goldstein [25]. The values of R_E that they reported are smaller than our own, perhaps because their coupling parameters were not quite optimal.[1] An energy stability analysis was carried out more recently for the IH3 configuration with no-slip boundaries [38], and those findings agree exactly with our own.

In the IH2 case, the energy stability eigenproblem can be solved exactly using long-wavelength asymptotics, much like the linear stability eigenproblem (cf. Sect. 2.1.3). This is possible because the infimum of expression (2.35) is reached as $k^2 \to 0$, an observation that has not been proven but has been confirmed numerically [15]. The asymptotic calculations, which are detailed in [15], give the exact expressions

$$R_E = \begin{cases} 2880\left(6\sqrt{35}-35\right) & \text{no-slip} \\ 360\left(9\sqrt{385}-175\right) & \text{free-slip top} \\ 1440\left(6\sqrt{35}-35\right) & \text{free-slip bottom} \\ 1440\left(8\sqrt{7}-21\right) & \text{free-slip} \end{cases} \qquad (2.36)$$

Numerical approximations of the above values appear in Table 2.3 above.

From the standpoint of scientific and engineering applications, the value of knowing R_E in IH convection is that we know convection cannot be sustained when $R < R_E$. When R lies between R_E and R_L, little is known about when convection can occur, apart from some instances of subcritical convection that have been computed in IH1 [42] and IH3 [35, 43]. This ambiguous regime between R_E and R_L is small in IH2 and IH3, and thus of not much practical importance, but it is much larger in IH1. In any event, we cannot claim to fully understand the static states until we know when subcritical convection is possible—that is, until we know the true value of R_c for every Pr. Lower bounds on R_c could be improved by looking beyond the energy method to find better Lyapunov functionals, and upper bounds could be improved by numerically computing steady states that exist in the subcritical regimes.

---

[1]What we call the IH1 case is designated in [25] by the parameter $Bi_0 = \infty$, and what we call the IH3 case is designated by the parameters $Bi_0 = 0$ and $Bi_1 = \infty$. Their Rayleigh numbers are converted to our scaling upon multiplication by 64.

2.3   Bounds Depending on the Rayleigh Number

Our main goal is to predict the parameter-dependence of integral quantities like $\langle wT \rangle$, $\delta\langle T \rangle$, and $\overline{T}_{\max}$. Much of this effort is equivalent to seeking the functions $N(R, Pr)$ and $\tilde{N}(R, Pr)$, where these Nusselt numbers are defined as in Tables 1.1 and 1.2. (Such functions are multivalued in general since multiple locally attracting solutions can coexist at a given set of parameters.) The stability analyses of Sects. 2.1 and 2.2 are useful because they give necessary and sufficient conditions for N and $\tilde{N}$ to equal unity. In particular, $R < R_E$ guarantees that both quantities equal unity, and $R > R_L$ guarantees that both are greater than unity. At large R, where convection is strong and complicated, exact expressions for $N(R, Pr)$ and $\tilde{N}(R, Pr)$ are not available. Instead, we seek to bound these quantities analytically.

The only parameter-dependent bounds that have been proven for RB or IH configurations can be stated as upper bounds on how quickly N or $\tilde{N}$, respectively, can grow as R is raised. Upper bounds on N have not been proven in IH convection but seem likely to hold (cf. Sect. 1.6.4). We cannot improve the lower bounds of unity since known techniques cannot distinguish realizable solutions from the unstable static states.

In this section we outline a proof of lower bounds on the mean temperature, $\delta\langle T \rangle := \langle T - \overline{T}_T \rangle$, in IH convection. (Recall that $\delta\langle T \rangle \equiv \langle T \rangle$ in IH1 and IH3 but not in IH2, and that lower bounds on $\delta\langle T \rangle$ are equivalent to upper bounds on $\tilde{N}$.) Our exposition combines existing results for IH1 [27] and IH2 [15] and a new result for IH3. The proof employs the background method [10, 12], which requires no assumptions beyond the governing equations. Like similar variational methods [2, 17, 18], the background method makes progress by relaxing the constraints on $\mathbf{u}$ and T. Instead of enforcing the full Boussinesq equations, we enforce only incompressibility, the boundary conditions, and a few integral relations that follow from the governing equations. This yields bounds that hold for an enlarged class of $\mathbf{u}$ and T that includes solutions of the Boussinesq equations.

Two main integral relations are typically enforced when the background method is applied to convection. They are called the power integrals and are derived by taking $\langle \mathbf{u} \cdot (1.7) \rangle$ and $\langle T \times (1.8) \rangle$ and integrating by parts to find

$$\langle |\nabla \mathbf{u}|^2 \rangle = R \langle wT \rangle \tag{2.37}$$

$$\langle |\nabla T|^2 \rangle = \begin{cases} 1 + \langle wT \rangle & \text{RB1} \\ \delta\overline{T} = 1 - \langle wT \rangle & \text{RB2, RB3} \\ \delta\langle T \rangle & \text{IH1, IH2, IH3.} \end{cases} \tag{2.38}$$

Time derivatives have vanished from the above relations in the infinite-time limit since the volume integrals of $|\mathbf{u}|$ and $|T|$ are bounded uniformly in time. This boundedness is proven as a by-product of the background-method analysis itself [10, 13]. The absence of Pr from the relaxed constraints on $\mathbf{u}$ and T precludes our analysis from producing bounds that depend on Pr.

In all six RB and IH configurations, the bounds that have been proven by the background method amount to bounds on the thermal dissipation, $\langle |\nabla T|^2 \rangle$, though they are often stated in terms of quantities like $\langle wT \rangle$ or $\delta \langle T \rangle$ that are related to $\langle |\nabla T|^2 \rangle$ by (2.38). The thermal dissipation is bounded above in RB1 and below in the other five cases. These results constitute upper bounds on N in RB convection and upper bounds on $\tilde{N}$ in IH convection.

2.3.1 Proof by the Background Method

We now prove for all three IH configurations that the dimensionless mean temperature, $\delta \langle T \rangle$, decays no faster than $R^{-1/3}$. This is equivalent to the *dimensional* mean temperature, $\delta \langle T \rangle \Delta$, growing with the rate of volumetric heating, H, no slower than $H^{2/3}$. We assume a no-slip top in the IH2 case but need not do so in the IH1 or IH3 cases.

To apply the background method, we decompose the temperature into a so-called background profile, $\tau(z)$, and the remaining part, $\Theta(\mathbf{x}, t)$:

$$T(\mathbf{x}, t) = \tau(z) + \Theta(\mathbf{x}, t). \tag{2.39}$$

The bound we obtain depends on the $\tau(z)$ we choose. The background profile does not generally solve the governing equations, in which case Θ does not evolve according to the fluctuation equations (2.1)–(2.3).

The $\tau(z)$ we choose must satisfy several conditions. First, it must be continuous. Second, it must satisfy the same boundary conditions as T since this lets Θ satisfy the corresponding homogenous conditions. In practice, however, only the fixed-temperature conditions on $\tau(z)$ need to be enforced. This is because fixed-flux conditions on $\tau(z)$ can be met by boundary layers whose influence vanishes as we send their thicknesses to zero [15]. These limiting bounds are the same as those reached by simply ignoring the fixed-flux conditions on $\tau(z)$, so we do the latter in our calculations. Finally, $\tau(z)$ must be chosen to make a particular quantity nonnegative, as explained below. We will see that for all admissible $\tau(z)$

$$\delta \langle T \rangle \geq 2 \langle \tau - \tau_T \rangle - \langle \tau'^2 \rangle. \tag{2.40}$$

We choose simple $\tau(z)$ that make our calculations analytically tractable, thereby yielding analytical bounds that are valid at all R. Optimizing $\tau(z)$ numerically at a given R would give a tighter bound (as in [32]), but the bound would apply only at that value of R.

To see where the inequality (2.40) comes from, and when it holds, we expand the power integral (2.38) for the IH cases using the decomposition (2.39) to find

$$\delta \langle T \rangle = \langle |\nabla T|^2 \rangle = \langle \tau'^2 \rangle + 2 \langle \tau' \Theta' \rangle + \langle |\nabla \Theta|^2 \rangle, \tag{2.41}$$

where primes denote z-derivatives. Our goal is to bound $\delta\langle T\rangle$ below. (We could equally well speak of bounding $\langle|\nabla T|^2\rangle$ below or bounding $\tilde{N}$ above.) The $\langle\tau'\Theta'\rangle$ term in the above expression is difficult to bound, so we eliminate it using a third and final integral relation. Integrating $\tau(z)$ against the temperature equation (1.8) gives the needed relation [27],

$$\langle\tau'\Theta'\rangle = \langle\tau - \tau_T\rangle - \langle\tau'^2\rangle + \langle\tau'w\Theta\rangle, \tag{2.42}$$

where the top temperature τ_T may be nonzero only in the IH2 case. Eliminating $\langle\tau'\Theta'\rangle$ from expression (2.41) gives

$$\delta\langle T\rangle = 2\langle\tau - \tau_T\rangle - \langle\tau'^2\rangle + \langle|\nabla\Theta|^2\rangle + 2\langle\tau'w\Theta\rangle. \tag{2.43}$$

From the above equality it follows that the lower bound (2.40) would hold if we could show $\langle|\nabla\Theta|^2\rangle + 2\langle\tau'w\Theta\rangle \geq 0$. This is an impossible task for arbitrary w, however, since the velocity enters only in the sign-indefinite term. Apparently, the temperature power integral (2.38) alone is not sufficiently constraining. We need the additional constraint of the velocity power integral (2.37), which tells us that $a\left(\frac{1}{R}\langle|\nabla\mathbf{u}|^2\rangle - \langle wT\rangle\right) = 0$ for any a. Adding this relation to (2.43) shows that the lower bound (2.40) on $\delta\langle T\rangle$ would follow from the nonnegativity of the quadratic functional

$$\mathcal{Q}[\mathbf{u},\Theta;\tau(z,R)] := \frac{a}{R}\langle|\nabla\mathbf{u}|^2\rangle + \langle|\nabla\Theta|^2\rangle + \langle(2\tau' - a)w\Theta\rangle. \tag{2.44}$$

We must choose a $\tau(z)$ for which we can verify that $\mathcal{Q} \geq 0$ for all admissible $\mathbf{u}$ and Θ.

More generally, the background method is carried out by finding an expression, equal to the quantity to be bounded, that takes the form $\mathcal{B} + \mathcal{Q}$, where $\mathcal{B}$ is a functional of the background field alone, while $\mathcal{Q}$ depends also on the other fields. The key idea is that $\mathcal{B}$ will be a lower bound when we can show that $\mathcal{Q}$ is nonnegative (or an upper bound when we can show that $\mathcal{Q}$ is nonpositive). In the present analysis, $\mathcal{B} := 2\langle\tau - \tau_T\rangle - \langle\tau'^2\rangle$, and $\mathcal{Q}$ is as defined in (2.44).

Two objectives compete in the choice of $\tau(z)$: making the lower bound (2.40) as large as possible, and maintaining the nonnegativity of $\mathcal{Q}$ that is needed for that bound to be valid. Here, we optimize $\tau(z)$ only among profiles consisting of two linear pieces. Such profiles can all be written in the ansatz

$$\tau(z) = \begin{cases} \left[\frac{b}{\delta} + \frac{a}{2}\left(\frac{1}{\delta} - 1\right)\right](1-z) & 1 - \delta \leq z \leq 1 \\ b + \frac{a}{2}z & 0 \leq z \leq 1 - \delta, \end{cases} \tag{2.45}$$

where the geometric meanings of parameters a, b, and δ are shown in Fig. 2.3. We will see that the top piece of $\tau(z)$ is a boundary layer whose thickness, δ, goes to zero as $R \to \infty$. The bottom piece of $\tau(z)$ has a slope that is half the value of the

Fig. 2.3 Schematic of the class of background profiles, $\tau(z)$, that we consider. The parameters a, b, and δ are optimized, within some constraints, to maximize the lower bounds on the mean temperature

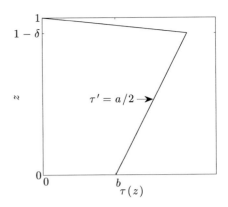

yet-unspecified constant a, a known trick [10, 27] for making the sign-indefinite term of $\mathscr{Q}$ vanish outside the boundary layer.

For our three-parameter family of background profiles (2.45), the lower bound (2.40) becomes

$$\delta\langle T\rangle \geq b(2-\delta) + \tfrac{a}{2}(1-\delta) - \left(\tfrac{a^2}{4} + ab\right)\left(\tfrac{1}{\delta} - 1\right) - \tfrac{b^2}{\delta}. \tag{2.46}$$

With a no-slip top in IH2 and any velocity conditions in IH1 or IH3, it can be shown that $\mathscr{Q} \geq 0$ is satisfied when δ is no larger than [15, 27]

$$\delta^4 = \begin{cases} 64\,aR^{-1} & \text{IH1, IH3} \\ 32\,aR^{-1} & \text{IH2.} \end{cases} \tag{2.47}$$

We choose this δ because the tightest bounds result from choosing δ as large as possible.

In the IH2 and IH3 cases, we are free to choose the parameters δ, a, and b to maximize the lower bound (2.46) subject to (2.47). In IH1, the lower boundary condition requires that $b = 0$, so we are free to choose only δ and a. This maximization is carried out for IH1 and IH2 in [27] and [15], respectively, and the procedure for IH3 is analogous. The resulting optimal parameters are

$$\delta^* = \begin{cases} 4R^{-1/3} \\ 12^{1/3}R^{-1/3} \\ 2\cdot 3^{1/3}R^{-1/3} \end{cases} \quad a^* = \begin{cases} 4R^{-1/3} \\ \frac{3\cdot 12^{1/3}}{8}R^{-1/3} \\ \frac{3^{4/3}}{4}R^{-1/3} \end{cases} \quad b^* = \begin{cases} 0 & \text{IH1} \\ \frac{5\cdot 12^{1/3}}{16}R^{-1/3} & \text{IH2} \\ \frac{5\cdot 3^{1/3}}{8}R^{-1/3} & \text{IH3,} \end{cases} \tag{2.48}$$

for which the lower bound (2.46) becomes

$$\delta\langle T\rangle \geq \begin{cases} R^{-1/3} - \qquad\qquad 4R^{-2/3} & \text{IH1} \\ \frac{9}{8}\left(\frac{3}{2}\right)^{1/3}R^{-1/3} - \frac{89}{64}\left(\frac{3}{2}\right)^{2/3}R^{-2/3} & \text{IH2} \\ \frac{3^{7/3}}{8}R^{-1/3} - \frac{89\cdot 3^{2/3}}{64}R^{-2/3} & \text{IH3.} \end{cases} \qquad (2.49)$$

At large R, the leading-order terms of the bounds dominate:

$$\delta\langle T\rangle \gtrsim \begin{cases} R^{-1/3} & \text{IH1} \\ 1.28\, R^{-1/3} & \text{IH2} \\ 1.62\, R^{-1/3} & \text{IH3.} \end{cases} \qquad (2.50)$$

When Lu et al. [27] proved the above bound for IH1, they also raised the prefactor from 1 to 1.09 by generalizing the ansatz of $\tau(z)$ to include a bottom boundary layer, although this required solving an algebraic equation numerically. Their proof carries through for the IH3 case also, so they in fact proved the asymptotic lower bound of $1.09\, R^{-1/3}$ for both IH1 and IH3. By dropping the condition $\tau(0) = 0$ in IH3, where it is not needed, we have raised the prefactor to 1.62. Optimizing $\tau(z)$ beyond our limited ansatz would lower the prefactors of the bounds, but results of numerically optimizing $\tau(z)$ in the RB1 case suggest that the scaling of the bounds would not change [32].

2.3.2 Similarities Between RB and IH Bounds

A main virtue of the way we have defined the Nusselt numbers N and $\tilde{N}$ and the diagnostic Rayleigh number Ra and $\widetilde{Ra}$ is that bounds for the various configurations all have the same scaling when expressed using these quantities. Recalling that the definitions (1.33) of $\tilde{N}$ are inversely proportional to $\delta\langle T\rangle$, and that $\widetilde{Ra} = R/\tilde{N}$ in IH convection, we see that the asymptotic bounds (2.50) on $\delta\langle T\rangle$ become

$$\tilde{N} \lesssim \begin{cases} 0.025\, \widetilde{Ra}^{1/2} & \text{IH1} \\ 0.132\, \widetilde{Ra}^{1/2} & \text{IH2} \\ 0.094\, \widetilde{Ra}^{1/2} & \text{IH3.} \end{cases} \qquad (2.51)$$

These upper bounds on $\tilde{N}$ appear in Table 2.1 at the start of this chapter, along with the best known bounds on N in the RB configurations. The RB bounds have different prefactors but the same exponent, scaling proportionally to $Ra^{1/2}$. The RB1 prefactor in Table 2.1 comes from the improvement on [10] by Plasting and Kerswell [32], who also showed that their bound could not be improved without additional constraints. The prefactor in the other two RB cases comes from [29], where the analysis was aimed at RB2 but carries through for RB3 also.

Upper bounds with an exponent of $1/2$ are the best available for three-dimensional convection in general, but bounds with smaller exponents have been proven in special cases. Here too, analogies hold between various configurations if results are stated in terms of our diagnostic parameters. When the boundaries are free-slip, and either $Pr = \infty$ or the flow is two-dimensional, upper bounds with exponents of $5/12$ have been proven for RB1 [47, 48] and IH1 [46, 48]. When $Pr = \infty$ with no-slip boundaries, the best known bounds on N scale like $Ra^{1/3}(\log\log Ra)^{1/3}$ in RB1 [30] and like $Ra^{1/3}(\log Ra)^{1/2}$ in RB2 and RB3 [49], and the best known bound on $\tilde{N}$ scale like $\widetilde{Ra}^{1/3}(\log\widetilde{Ra})^{1/3}$ in IH1 [46]. Bounds with exponents smaller than $1/2$ are yet to be reported for the other RB or IH configurations.

Now that we have seen how the background method works, we can understand why it is challenging in the IH cases to prove upper bounds on $\langle wT \rangle$. This quantity is related to $\langle |\nabla \mathbf{u}|^2 \rangle$ by the velocity power integral (2.37) but is not is not related a priori to $\langle |\nabla T|^2 \rangle$ in IH convection. However, the parameter-dependent bounds that have been proven for convective models all amount to bounds on $\langle |\nabla T|^2 \rangle$ and rely on a background decomposition of the temperature field. Bounding $\langle |\nabla \mathbf{u}|^2 \rangle$ instead suggests a background decomposition of the velocity field, which has been carried out for shear flows (e.g., in [9]) but not for convection.

References

1. Ames, K.A., Straughan, B.: Penetrative convection in fluid layers with internal heat sources. Acta Mech. **85**, 137–148 (1990)
2. Busse, F.H.: On Howard's upper bound for heat transport by turbulent convection. J. Fluid Mech. **37**(3), 457–477 (1969)
3. Busse, F.H.: Remarks on the critical value $P_c = 0.25$ of the Prandtl number for internally heated convection found by Tveitereid and Palm. Eur. J. Mech. B/Fluids **47**, 32–34 (2014)
4. Chandrasekhar, S.: Hydrodynamic and Hydromagnetic Stability. Dover, New York (1981)
5. Chapman, C.J., Childress, S., Proctor, M.R.E.: Long wavelength thermal convection between non-conducting boundaries. Earth Planet. Sci. Lett. **51**, 362–369 (1980)
6. Chapman, C.J., Proctor, M.R.E.: Nonlinear Rayleigh–Bénard convection between poorly conducting boundaries. J. Fluid Mech. **101**(04), 759–782 (1980)
7. Chernyshenko, S.I., Goulart, P., Huang, D., Papachristodoulou, A.: Polynomial sum of squares in fluid dynamics: a review with a look ahead. Philos. Trans. R. Soc. A **372**, 20130350 (2014)
8. Childress, S., Spiegel, E.A.: Pattern formation in a suspension of swimming microorganisms: nonlinear aspects. In: Givoli, D., Grote, M.J., Papanicolaou, G.C. (eds.) A Celebr. Math. Model. Kluwer Academic Publishers, New York (2004)
9. Constantin, P., Doering, C.R.: Variational bounds on energy dissipation in incompressible flows: Shear flow. Phys. Rev. E **49**(5), 4087–4099 (1994)
10. Constantin, P., Doering, C.R.: Variational bounds on energy dissipation in incompressible flows. III. Convection. Phys. Rev. E **53**(6), 5957–5981 (1996)
11. Debler, W.R.: The onset of laminar natural convection in a fluid with homogenously distributed heat sources. Ph.D. thesis, University of Michigan (1959)
12. Doering, C.R., Constantin, P.: Energy dissipation in shear driven turbulence. Phys. Rev. Lett. **69**(11), 1648–1651 (1992)

13. Doering, C.R., Gibbon, J.D.: Applied Analysis of the Navier-Stokes Equations. Cambridge University Press, Cambridge (1995)
14. Galdi, G.P., Straughan, B.: Exchange of stabilities, symmetry, and nonlinear stability. Arch. Ration. Mech. Anal. **89**(3), 211–228 (1985)
15. Goluskin, D.: Internally heated convection beneath a poor conductor. J. Fluid Mech. **771**, 36–56 (2015)
16. Herron, I.H.: On the principle of exchange of stabilities in Rayleigh-Bénard convection, II - No-slip boundary conditions. Ann. dell'Università di Ferrara **IL**, 169–182 (2003)
17. Howard, L.N.: Heat transport by turbulent convection. J. Fluid Mech. **17**(3), 405–432 (1963)
18. Howard, L.N.: Bounds on flow quantities. Annu. Rev. Fluid Mech. **4**, 473–494 (1972)
19. Ishiwatari, M., Takehiro, S.I., Hayashi, Y.Y.: The effects of thermal conditions on the cell sizes of two-dimensional convection. J. Fluid Mech. **281**, 33–50 (1994)
20. Jeffreys, H.: Some cases of instability in fluid motion. Proc. R. Soc. A **118**, 195–208 (1928)
21. Joseph, D.D.: On the stability of the Boussinesq equations. Arch. Ration. Mech. Anal. **20**(1), 59–71 (1965)
22. Joseph, D.D.: Stability of Fluid Motions I-II. Springer, New York (1976)
23. Kaiser, R., Tilgner, A., Von Wahl, W.: A generalized energy functional for plane Couette flow. SIAM J. Math. Anal. **37**(2), 438–454 (2005)
24. Kulacki, F.A., Goldstein, R.J.: Hydrodynamic instability in fluid layers with uniform volumetric energy sources. Appl. Sci. Res. **31**(2), 81–109 (1975)
25. Kulacki, F.A., Nagle, M.E.: Natural convection in a horizontal fluid layer with volumetric energy sources. J. Heat Transfer **97**, 204–211 (1975)
26. Low, A.R.: On the criterion for stability of a layer of viscous fluid heated from below. Proc. R. Soc. A **125**(796), 180–195 (1929)
27. Lu, L., Doering, C.R., Busse, F.H.: Bounds on convection driven by internal heating. J. Math. Phys. **45**(7), 2967–2986 (2004)
28. McKenzie, D.P., Roberts, J.M., Weiss, N.O.: Convection in the earth's mantle: towards a numerical simulation. J. Fluid Mech. **62**(3), 465–538 (1974)
29. Otero, J., Wittenberg, R.W., Worthing, R.A., Doering, C.R.: Bounds on Rayleigh-Bénard convection with an imposed heat flux. J. Fluid Mech. **473**, 191–199 (2002)
30. Otto, F., Seis, C.: Rayleigh-Bénard convection: Improved bounds on the Nusselt number. J. Math. Phys. **52**(8), 083702 (2011)
31. Pellew, A., Southwell, R.V.: On maintained convective motion in a fluid heated from below. Proc. R. Soc. A **176**, 312–343 (1940)
32. Plasting, S.C., Kerswell, R.R.: Improved upper bound on the energy dissipation rate in plane Couette flow: the full solution to Busse's problem and the Constantin-Doering-Hopf problem with one-dimensional background field. J. Fluid Mech. **477**, 363–379 (2003)
33. Rayleigh, Lord: On convection currents in a horizontal layer of fluid, when the higher temperature is on the under side. Philos. Mag. **32**(192), 529–546 (1916)
34. Roberts, P.H.: Convection in horizontal layers with internal heat generation. Theory. J. Fluid Mech. **30**(01), 33–49 (1967)
35. Schwiderski, E.W.: Bifurcation of convection in internally heated fluid layers. Phys. Fluids **15**, 1882–1898 (1972)
36. Serrin, J.: On the stability of viscous fluid motions. Arch. Ration. Mech. Anal. **3**(1), 1–13 (1959)
37. Sparrow, E.M., Goldstein, R.J., Jonsson, V.K.: Thermal instability in a horizontal fluid layer: effect of boundary conditions and non-linear temperature profile. J. Fluid Mech. **18**(04), 513–528 (1964)
38. Straughan, B.: Continuous dependence on the heat source and non-linear stability for convection with internal heat generation. Math. Methods Appl. Sci. **13**, 373–383 (1990)
39. Straughan, B.: The Energy Method, Stability, and Nonlinear Convection, 2 edn. Springer, New York (2004)
40. Thirlby, R.: Convection in an internally heated layer. J. Fluid Mech. **44**(04), 673–693 (1970)
41. Trefethen, L.N.: Spectral Methods in MATLAB. SIAM, Philadelphia (2000)

42. Tveitereid, M.: Thermal convection in a horizontal fluid layer with internal heat sources. Int. J. Heat Mass Transfer **21**, 335–339 (1978)
43. Tveitereid, M., Palm, E.: Convection due to internal heat sources. J. Fluid Mech. **76**(03), 481 (1976)
44. Veronis, G.: Penetrative convection. Astrophys. J. **137**, 641–663 (1962)
45. Watson, P.M.: Classical cellular convection with a spatial heat source. J. Fluid Mech. **32**, 399 (1968)
46. Whitehead, J.P., Doering, C.R.: Internal heating driven convection at infinite Prandtl number. J. Math. Phys. **52**(9), 093101 (2011)
47. Whitehead, J.P., Doering, C.R.: Ultimate state of two-dimensional Rayleigh-Bénard convection between free-slip fixed-temperature boundaries. Phys. Rev. Lett. **106**(24), 244501 (2011)
48. Whitehead, J.P., Doering, C.R.: Rigid bounds on heat transport by a fluid between slippery boundaries. J. Fluid Mech. **707**, 241–259 (2012)
49. Whitehead, J.P., Wittenberg, R.W.: A rigorous bound on the vertical transport of heat in Rayleigh-Bénard convection at infinite Prandtl number with mixed thermal boundary conditions. J. Math. Phys. **55**(9), 093104 (2014)

Chapter 3
Internally Heated Convection Experiments and Simulations

Abstract Laboratory experiments and numerical simulations studying internally heated convection are reviewed. The emphasis is on quantitative results, especially integral quantities important to heat transport and their dependence on the Rayleigh number, which is proportional to the heating rate. For all experiments and three-dimensional simulations, the various measures of fluid temperature can be fit to powers of the rate of volumetric heating. The exponents of these fits range from 0.75 to 0.77 when the bottom is insulating, and they range from 0.78 to 0.82 when the top and bottom are fixed at equal temperatures. In the latter configuration, the fraction of internally produced heat flowing outward across the bottom boundary falls quite slowly as heating is strengthened. When this fraction is fit to a power of the heating rate, the fit exponents lie between -0.049 and -0.099.

The first two chapters have summarized features of heat transport in IH convection and RB convection that can be ascertained analytically from the Boussinesq equations. In this final chapter we summarize findings on IH convection from physical and computational experiments. Analogous results for RB convection are described only minimally, as the experimental literature on RB convection is vast and has been reviewed elsewhere (e.g., [1, 18, 26, 52, 68]).

Precise laboratory experiments on IH convections are inherently more difficult to carry out than similar experiments on RB convection. Both require maintaining the chosen thermal boundary conditions, but IH experiments also require producing heat internally in a controlled way. If our simple models are to apply, the heat production should be constant and uniform. In most experiments, the internal heating has been achieved by Joule heating, where the working fluid is an electrolytic solution that is heated by passing current through it. Two sets of experiments [50, 59] used a different method, wherein heating elements were distributed throughout the domain. Although neither method heats uniformly, it is possible that rapid mixing by strong convection limits the influence of non-uniformity. This is supported by the fairly good agreement between non-uniformly heated experiments and uniformly heated simulations.

Numerical simulations of IH convection avoid unknown variations in heating rate or material properties. However, the majority of numerical studies were carried out several decades ago and were limited to 2D and fairly small R. The larger values

© Springer International Publishing Switzerland 2016

D. Goluskin, *Internally Heated Convection and Rayleigh-Bénard Convection*,
SpringerBriefs in Applied Sciences and Technology,
DOI 10.1007/978-3-319-23941-5_3

of R accessible with modern computers have been simulated only a few times, and much of the parameter space that could now be reached has yet to be explored.

Experimental findings before 1985 are collected in the review of Kulacki and Richards [49], who discuss findings on IH1, IH3, and some similar configurations. The slightly later review of Cheung and Chawla [17] adds various scaling arguments for heat transport. Nourgaliev et al. [55] summarize heat fluxes in these same early experiments, as well as in experiments with curved geometries and cooled side walls.

A number of experiments have examined heat transport quantitatively, and a number of others have focused on qualitative pattern formation near the onset of convection. Here we cite studies of both types but focus on quantitative findings, and we restrict ourselves to experiments that closely resemble one of the IH config-urations defined in Fig. 1.1. Convection with internal heating has been studied also with various complications that we do not confront, such as cooled side walls [3–6, 14, 21, 25, 34, 51, 55, 67, 71], non-uniform heating [45, 60, 73, 76], self-gravitating spheres [8, 9, 37, 61–63], and hybrid configurations driven both internally and by the boundary conditions [2, 7, 12, 19, 31, 35, 42–44, 54, 69, 70, 74, 83].

Section 3.1 addresses IH3, the last of the three IH configurations in Fig. 1.1. Section 3.2 addresses IH1, which is in some ways more complicated than IH3. We are not aware of any heat transport findings on the IH2 configuration, though 2D simulations have been carried out to study scale selection [33, 38]. Section 3.3 suggests directions for future work.

3.1   The IH3 Configuration

The internally heated configuration we call IH3, which is bounded above by a perfect conductor and below by a perfect insulator, has been the subject of numerous laboratory experiments [20, 24, 46, 48, 50, 58, 59, 66, 75, 77, 79], as well as com-putational studies both in 2D [22, 38, 54, 78] and in 3D [10, 11, 35, 36, 65, 78, 81]. Many of these investigations have focused on pattern formation and scale selection, which we do not discuss here. Our interest is in quantities relevant to heat transport, including the mean vertical temperature profile, $\overline{T}(z)$, and the mean temperature difference between the boundaries, $\delta\overline{T}$. No data are available on the mean fluid temperature, $\delta\langle T \rangle$.

Computational studies of IH3 convection that report $\overline{T}(z)$ or $\delta\overline{T}$ are all several decades old. Most are limited to the steady states that are stable at modest R [54, 65, 78, 81]. At larger R, unsteady 2D simulations have been carried out using a turbulence closure model [23] and by direct numerical simulation (DNS) [22], although some runs in the latter study seem under-resolved. As far as we know, unsteady IH3 convection has not been simulated in 3D. The largest R that has been reached in 2D DNS of IH3 [22] could be greatly exceeded in 3D DNS on modern parallel computers.

Laboratory experiments, most of which were carried out in the 1970s, furnish nearly everything we know about $\overline{T}(z)$ and $\delta\overline{T}$ in IH3 convection at large R [24, 46, 48, 50, 58, 59]. These findings are subject to the uncertainties inherent to IH experiments, so there is cause to repeat them numerically. The largest R reached in past laboratory experiments of IH3 could now be approached by 3D DNS, albeit in a smaller spatial domain.

3.1.1 Temperature Profiles

Several authors have reported mean vertical temperature profiles. It is simple to obtain $\overline{T}(z)$ in numerical studies by averaging steady flows horizontally [54, 78, 81] or averaging unsteady flows both horizontally and temporally [22, 23]. In laboratory experiments, vertical profiles have been obtained by measuring temperatures at fixed points and averaging only over time [58]. If the flow is horizontally isotropic in a statistical sense, and time averages are sufficiently long, then the same mean profile would be obtained whether or not horizontal averages are also taken. This is generally expected to be true at large R when side walls are absent or negligible. Some transient profiles have been reported also [46, 48], but these do not bear directly on the infinite-time averages we seek.

Figure 3.1 shows $\overline{T}(z)$ profiles for 2D steady states computed by Thirlby [78] for $Pr = 6.8$ and relatively small R. As R is raised and convection strengthens, the dimensionless temperature decreases, and the interior becomes closer to isothermal. When convection is sufficiently strong, the maximum value of $\overline{T}(z)$ occurs inside the layer, rather than at the bottom boundary. At still larger R, where convection is stronger and unsteady, the experimentally measured profiles of Ralph and Roberts [58] follow similar trends but are closer to isothermal in the interior, lacking the pronounced temperature inversion found in the steady states of Fig. 3.1. The

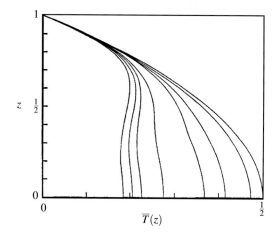

Fig. 3.1 Numerically computed mean temperature profiles, $\overline{T}(z)$, for steady 2D convection between no-slip boundaries. The Prandtl number is 6.8. The rightmost profile is that of the static state. The others, from *right* to *left*, are for $R = 10^3 \times (3, 4, 5, 10, 20, 30, 52)$. The figure is adapted from Fig. 2 of Thirlby [78]

unsteady motions responsible for homogenizing temperature outside the thermal boundary layer are evident in the IH3 temperature field of Fig. 1.2c: plumes emerge from the unstably stratified upper boundary layer and strongly mix fluid in the rest of the domain.

3.1.2 Mean Temperature Differences

The difference by which the dimensionless temperature at any point exceeds its average value at the top boundary, $T - \overline{T}_T$, tends to decrease as IH convection strengthens. This fact underlies the two related but distinct measures of convective strength discussed in Sect. 1.6: $\delta\langle T\rangle$, which is the average of $T - \overline{T}_T$ over time and the entire volume, and $\delta\overline{T}$, which is its average over time and the bottom boundary. (Recall that $\delta\overline{T}$ is also the mean vertical conduction, and in IH3 it is tied to the mean vertical convection, $\langle wT\rangle$, by the relation $\delta\overline{T} + \langle wT\rangle = 1/2$.) Both $\delta\overline{T}$ and $\delta\langle T\rangle$ are maximal in the static state and most likely approach zero in convective flows as $R \to \infty$.

A primary question for experimentalists is *how quickly* $\delta\overline{T}$ and $\delta\langle T\rangle$ fall as R is raised, along with how this answer is affected by the velocity boundary conditions, Prandtl number, and geometry. If $\delta\overline{T}$ and $\delta\langle T\rangle$ vary approximately as powers of R when other parameters are held constant, data can be captured by fits of the form

$$\delta\overline{T} \sim aR^{-\alpha} \qquad\qquad \delta\langle T\rangle \sim bR^{-\beta}. \qquad (3.1)$$

In the IH3 case, the Nusselt numbers and diagnostic Rayleigh numbers defined in Sects. 1.6.4 and 1.6.5 are

$$N = 1/2\delta\overline{T} \qquad\qquad Ra = R/N \qquad (3.2)$$

$$\tilde{N} = 1/3\delta\langle T\rangle \qquad\qquad \widetilde{Ra} = R/\tilde{N}. \qquad (3.3)$$

Various authors have considered quantities like N in past studies of IH3, and Fiedler and Wille [24] considered both N and Ra together. Restated in terms of the diagnostic variables, the fits of expression (3.1) become

$$N \sim cRa^\gamma \qquad\qquad \tilde{N} \sim d\widetilde{Ra}^\delta, \qquad (3.4)$$

were $\gamma = \alpha/(1-\alpha)$, $\delta = \beta/(1-\beta)$, $c = (2a)^{-1/(1-\alpha)}$, and $d = (3b)^{-1/(1-\beta)}$.

The mean temperature difference $\delta\overline{T}$ has been measured in a number of experiments. Table 3.1 summarizes past fits of the form $\delta\overline{T} \sim aR^{-\alpha}$, along with their corresponding re-expressions as fits of the form $N \sim cRa^\gamma$. Ranges of Pr and R are also given. The stated ranges of R are those over which the data have been fit. The Prandtl number would ideally be held constant as R is changed, but slight variations are unavoidable in the laboratory. Numerical studies do not suffer from

Table 3.1 Summary of IH3 experiments and simulations reporting approximate power-law dependence of $\delta\overline{T}$ on R

	Pr	R	$\delta\overline{T}$ fit	N fit
Laboratory experiments				
Fiedler and Wille [24]	6–7	10^4–10^7	$1.90\,R^{-0.228}$	$0.177\,Ra^{0.295}$
Ralph and Roberts [58]	6–7	$2.3\cdot10^5$–$6.0\cdot10^9$	$2.62\,R^{-0.25}$	$0.110\,Ra^{0.33}$
Kulacki and Nagle [48]	6.2–6.6	$1.5\cdot10^5$–$2.5\cdot10^9$	$3.28\,R^{-0.239}$	$0.0845\,Ra^{0.314}$
Kulacki and Emara [46]	2.7–6.9	$1.89\cdot10^3$–$2.17\cdot10^{12}$	$2.53\,R^{-0.227}$	$0.123\,Ra^{0.294}$
Ralph et al. [59]	6–7	10^9–$7\cdot10^9$	$a\,R^{-0.24}$	$c\,Ra^{0.32}$
Lee et al. [50]	0.71–0.74	$9.9\cdot10^9$–$3.3\cdot10^{11}$	$2.84\,R^{-0.247}$	$0.0996\,Ra^{0.328}$
Simulations (2D DNS)				
Mckenzie et al. [54] (free-slip, steady)	∞	$1.2\cdot10^4$–$7.0\cdot10^5$	$a\,R^{-0.26}$	$c\,Ra^{0.35}$
Emara and Kulacki [22] (free-slip top)	6.5	$5\cdot10^4$–$5\cdot10^8$	$1.07\,R^{-0.182}$	$0.397\,Ra^{0.222}$
Emara and Kulacki [22] (no-slip)	6.5	$5\cdot10^3$–$5\cdot10^8$	$2.38\,R^{-0.223}$	$0.134\,Ra^{0.287}$
Olwi [56] (no-slip, steady)	6.5	10^4–10^8	$3.07\,R^{-0.255}$	$0.0876\,Ra^{0.342}$

Internal heating was achieved by electric current in the first four experiments and by heating elements in the last two. The Prandtl number range 6–7 is an estimate for experiments that used aqueous solutions but did not report Pr measurements [24, 58, 59]

this uncertainty, but the simulation results in Table 3.1 nonetheless must be regarded with care since they all are 2D and seem to be somewhat under-resolved at larger R.

The decay rates of $\delta\overline{T}$ reported for the six laboratory experiments in Table 3.1 fall between $\alpha = 0.227$ and $\alpha = 0.25$. This means that the dimensional temperature difference between the boundaries, $\delta\overline{T}\Delta$, *grows* with the volumetric heating at rates between $H^{0.75}$ and $H^{0.773}$.

When the $\delta\overline{T}$ fits in Table 3.1 are restated in the form $N \sim c\,Ra^\gamma$, the exponents fall between $\gamma = 0.294$ and $\gamma = 0.33$. This range agrees very well with the analogous range of γ measured in RB1 experiments, where fits still take the form $N \sim c\,Ra^\gamma$, but with $N := 1 + \langle wT\rangle$ and $Ra := R$ (cf. Sect. 1.6.4). The RB1 exponents summarized in Table 1 of [29] lie between 0.25 and 0.33, excluding the very small Pr values for which corresponding IH3 data are unavailable. Exponents larger than 0.33 have sometimes been measured in RB1 experiments at very large R [13, 32], but no IH3 experiments have reached such R values. The similarity between measured values of γ in RB1 and IH3 is one of the analogies brought out by our chosen definitions of N and Ra.

No data have been reported on the volume-averaged quantity $\delta\langle T\rangle$, so we cannot say exactly what exponents would emerge from fits of the form $\delta\langle T\rangle \sim b\,R^{-\beta}$ or $\widetilde{N} \sim d\,Ra^{\widetilde{\delta}}$. We can reasonably estimate the exponents, however, since the temperature profiles that have been reported are close to isothermal outside their boundary layers. This suggests that the values of $\delta\langle T\rangle$ and $\delta\overline{T}$ become ever closer as $R \to \infty$, in which case $\alpha \approx \beta$ and $\gamma \approx \delta$ for sufficiently large R. This speculation remains to be tested since volume averages like $\delta\langle T \rangle$ are difficult to measure in the laboratory. They are easy to extract from simulations, however, and we hope that future numerical studies will report $\delta\langle T \rangle$.

Whereas we have data on $\overline{\delta T}$ but not on $\delta\langle T \rangle$—or, equivalently, on N but not on $\tilde{N}$—the state of affairs for analytical bounds is just the opposite. We have conjectured in Chap. 1, but have not proven, than N obeys an upper bound of the form $c\,Ra^{1/2}$. The experimental exponents β in Table 3.1 are all smaller than 1/2 and thus consistent with this conjecture. On the other hand, we *have* proven in Chap. 2 that $\tilde{N}$ can grow no faster than $0.093\,\widetilde{Ra}^{1/2}$, but no data on $\delta\langle T \rangle$ have been reported for the IH3 configuration.

3.2 The IH1 Configuration

The internally heated configuration we call IH1, which is bounded above and below by perfect conductors of equal temperature, has been studied in the laboratory [39–41, 47, 50, 53, 59], as well as numerically both in 2D [22, 27, 53, 57, 74, 80] and in 3D [28, 30, 84]. Almost all of these studies have reported quantitatively on heat transport in some way.

Numerical computations of IH1 include both steady states [57, 74, 80] and DNS. Whereas DNS of the IH3 configuration has been limited to a single 2D study, DNS of the IH1 configuration has been carried out up to fairly large R in both 2D [22, 27] and 3D [28, 30, 84].

3.2.1 Temperature Profiles

Mean vertical temperature profiles have been reported in a number of studies. Numerical studies provide profiles, $\overline{T}(z)$, that are averaged horizontally and, if the simulations are unsteady, over time as well [22, 27, 28, 30, 53, 57, 74, 84]. In the laboratory, profiles measured pointwise by temperature probes are averaged only over time [50, 59], while profiles gleaned from interferograms are instantaneous but effectively averaged over a horizontal direction [47, 53].

Figure 3.2a shows $\overline{T}(z)$ profiles from the 3D DNS data of Goluskin and van der Poel [28]. As in the IH3 configuration, raising R strengthens convection, which decreases the dimensionless temperature and brings the interior closer to isothermality. When R is large enough for thermal boundary layers to be discernible, the top boundary layer is visibly thinner than the bottom one. This reflects the up-down asymmetry of heat fluxes; more of the produced heat flows outward across the top boundary than across the bottom one, as quantified in the next subsection. The same basic features are evident in Fig. 3.2b, which shows an interferogram

Fig. 3.2 (**a**) Mean temperature profiles, $\overline{T}(z)$, from the 3D DNS of Goluskin and van der Poel [28] for a fluid with $Pr = 1$ between no-slip boundaries. The rightmost profile is that of the static state. The others, from *right* to *left*, are for $R = 10^6$, 10^7, 10^8, 10^9, and 10^{10}. (**b**) An interferogram from the experiments of Kulacki and Goldstein [47] with $R = 1.5 \cdot 10^5$ and $Pr = 5.8$. Any curve of constant color acts approximately as a graph of instantaneous temperature averaged over one horizontal direction

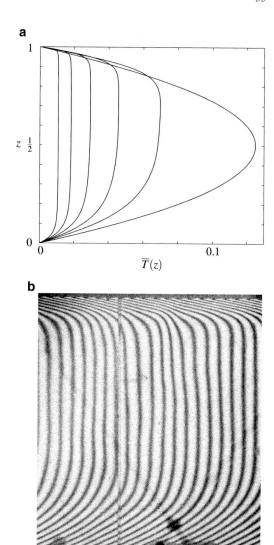

from the experiments of Kulacki and Goldstein [47]. The interferogram measures horizontally averaged optical properties of the fluid that vary with its temperature, and a line of constant color can be interpreted as a temperature profile.

The IH1 configuration stands out from the other RB and IH models we have discussed in that there is a stably stratified thermal boundary layer. The configuration thus provides a simple instance of *penetrative convection*, wherein buoyancy forces in an unstably stratified region drive motions that penetrate into a stably stratified region. The temperature field of Fig. 1.2b reflects the dissimilarity between the unstably stratified upper boundary layer and the stably stratified lower one. Mixing of the cold upper layer with the warmer interior is accomplished by buoyantly driven cold plumes. At large R, the cold *lower* layer also can mix with the warmer interior. This mixing is driven by shear forces, rather than by buoyancy, and it occurs when the interior turbulence pulls cold eddies off the bottom boundary layer.

3.2.2 Maximum Temperatures, Mean Temperatures, and Asymmetry

The mean fluid temperature, $\delta\langle T\rangle$, behaves in IH1 convection much like it does in IH3 convection, assuming its maximum value in the static state and falling as R is raised. On the other hand, the mean temperature change between the boundaries, $\delta\overline{T}$, differs completely between the two configurations. Whereas in IH3 $\delta\overline{T}$ behaves rather like $\delta\langle T\rangle$, in IH1 it is identically zero. The role that $\delta\overline{T}$ plays in IH3 is instead approximated in IH1 by $\overline{T}_{\max}$, the maximum value that $\overline{T}(z)$ assumes over the layer. Whereas $\delta\overline{T}$ equals the mean upward conduction across the entire layer, $\overline{T}_{\max}$ captures the mean *outward* conduction, as described in Sect. 1.6.4.1. The quantity $\overline{T}_{\max}$ has been reported in many studies of IH1 since it is easier to estimate in the laboratory than $\delta\langle T\rangle$. However, $\overline{T}_{\max}$ does not arise as easily as $\delta\langle T\rangle$ in analytical expressions.

In addition to $\delta\langle T\rangle$ and $\overline{T}_{\max}$, IH1 convection is naturally characterized by the extent to which the flow creates asymmetry between upward and downward heat fluxes. This asymmetry can be simply conveyed by the mean fractions of produced heat leaving across the top or bottom boundaries—$\mathscr{F}_T$ or $\mathscr{F}_B$. As described in Sect. 1.6.2, these fractions are related to the dimensionless convective flux, $\langle wT\rangle$, by

$$\mathscr{F}_T = \tfrac{1}{2} + \langle wT\rangle \qquad\qquad \mathscr{F}_B = \tfrac{1}{2} - \langle wT\rangle. \qquad (3.5)$$

One can equivalently speak in terms of $\langle wT\rangle$, $\mathscr{F}_T$, or $\mathscr{F}_B$. Here we focus on $\mathscr{F}_B$ because it comes the closest to having a power-law dependence on R as in the regimes studied.

Despite their simplicity and analytical attractiveness, neither $\delta\langle T\rangle$ nor $\mathscr{F}_B$ has received much attention, although both quantities have been mentioned. Most authors have instead spoken in terms of top and bottom Nusselt numbers, here called N_T and N_B. The most common definitions of these numbers are

$$N_T := \frac{\mathscr{F}_T}{\overline{T}_{\max}} \qquad\qquad N_B := \frac{\mathscr{F}_B}{\overline{T}_{\max}}. \qquad (3.6)$$

The above expressions are not normalized to be unity in the static state; instead, both are equal to 4. Data on N_T and N_B are typically fit to powers of R.

To keep measures of asymmetry and temperature as separate as possible, we prefer not to examine N_T and N_B. Instead, we use $\mathscr{F}_B$ as a measure of asymmetry and use $\overline{T}_{\max}$ and $\delta\langle T\rangle$—or their inverses, N and $\tilde{N}$—as measures of temperature. One undesirable feature of N_B is that it can initially drop below its static value as R

is raised since $\mathscr{F}_B$ can initially fall faster than $\overline{T}_{\max}$. For instance, this occurs in the data of Kulacki and Goldstein [47]. Such behavior prevents N_B from being well fit by a power of R near onset, and it is unlike the behavior of the RB Nusselt number, which cannot be smaller than its static value. Another disadvantage of using N_T and N_B is that their R-dependence differs only in regimes where $\mathscr{F}_B$ is changing significantly. If the decay of $\mathscr{F}_B$ stops, as in the 2D simulations of Goluskin and Spiegel [27], then N_T and N_B will both be dominated by the scaling of $1/\overline{T}_{\max}$, and slight changes in the asymmetry will not be captured well.

We would like to summarize past data on R-dependence with fits of the form

$$\overline{T}_{\max} \sim a R^{-\alpha} \qquad \delta\langle T \rangle \sim b R^{-\beta} \qquad \mathscr{F}_B \sim e R^{-\varepsilon}. \qquad (3.7)$$

Fits of the above form have been reported for all three quantities in [28] and for $\delta\langle T \rangle$ in [27]. In two other studies where the original data are available to us [47, 84], we have calculated fits to $\overline{T}_{\max}$ and $\mathscr{F}_B$. For the remaining studies, only fits to N_T and N_B are available. In these cases, we use the relations

$$\overline{T}_{\max} = \frac{1}{N_T + N_B} \qquad \mathscr{F}_B = \frac{N_B}{N_T + N_B}. \qquad (3.8)$$

The reported power-law fits to N_T and N_B define curves for $\overline{T}_{\max}$ and $\mathscr{F}_B$ that are not pure powers of R, so we have re-fit pure power laws to the latter curves.

Table 3.2 summarizes power-law fits to the R-dependence of $\overline{T}_{\max}$, $\delta\langle T \rangle$, and $\mathscr{F}_B$. The fits to $\overline{T}_{\max}$ are also stated in terms of N and Ra, and the fits to $\delta\langle T \rangle$ are also stated in terms of $\tilde{N}$ and $\widetilde{Ra}$. For the IH1 configuration, we have defined these diagnostic quantities in Sects. 1.6.4 and 1.6.5 as

$$N = 1/8\overline{T}_{\max} \qquad Ra = R/N \qquad (3.9)$$

$$\tilde{N} = 1/12\,\delta\langle T \rangle \qquad \widetilde{Ra} = R/\tilde{N}. \qquad (3.10)$$

The fits (3.7) to $\overline{T}_{\max}$ and $\delta\langle T \rangle$ imply fits to and N and $\tilde{N}$ of the form (3.4), where $\gamma = \alpha/(1-\alpha)$, $\delta = \beta/(1-\beta)$, $c = (8a)^{-1/(1-\alpha)}$, and $d = (12b)^{-1/(1-\beta)}$.

3.2.2.1 Maximum Temperatures

Fits of the form $\overline{T}_{\max} \sim a R^{-\alpha}$ are shown in Table 3.2. In all laboratory experiments and all simulations with no-slip boundaries, the exponent α lies between 0.180 and 0.224. This means that the dimensional maximum temperature, $\overline{T}_{\max}\Delta$, grows with the volumetric heating at rates between $H^{0.776}$ and $H^{0.820}$. In the sole study for which both $\overline{T}_{\max}$ and $\delta\langle T \rangle$ are reported [28], the decay of $\overline{T}_{\max}$ is slightly faster than the decay of $\delta\langle T \rangle$, the fit exponents being $\alpha = 0.217$ and $\beta = 0.204$, respectively. This makes sense since $\overline{T}_{\max}$ initially must "catch up" to $\delta\langle T \rangle$ as the temperature profile flattens. When the $\overline{T}_{\max}$ fits are restated in the form $N \sim c Ra^{\gamma}$, the exponents

Table 3.2 Summary of IH1 experiments and simulations reporting approximate power-law dependence of $\overline{T}_{\max}$, $\delta\overline{T}$, or $\mathscr{F}_B$ on R

	Pr	R	$\overline{T}_{\max}$ fit	N fit	$\delta\langle T\rangle$ fit	$\tilde{N}$ fit	$\mathscr{F}_B$ fit
Laboratory experiments							
Kulacki and Goldstein [47]	5.7–6.3	$R_L - 2.4\cdot10^7$	$1.71R^{-0.180}$	$0.0958Ra^{0.219}$			$1.21R^{-0.0848}$
Jahn and Reineke [39, 53]	≈ 7	10^5–10^9	$1.96R^{-0.194}$	$0.0778Ra^{0.240}$			$1.36R^{-0.0988}$
Ralph et al. [59]	6–7	$3.7\cdot10^8 - 1.1\cdot10^{12}$	$5.39R^{-0.224}$	$0.0191Ra^{0.289}$			$0.692R^{-0.0494}$
Lee et al. [50]	0.71–0.74	$1.1\cdot10^{10} - 3.7\cdot10^{11}$	$3.86R^{-0.209}$	$0.0315Ra^{0.264}$			$2.48R^{-0.0947}$
Simulations (3D DNS)							
Wörner et al. [84]	7	10^5–10^8	$1.86R^{-0.186}$	$0.0847Ra^{0.229}$			$1.16R^{-0.0845}$
Goluskin and van der Poel [28]	1	$5\cdot10^7$–$2\cdot10^{10}$	$1.62R^{-0.217}$	$0.0379Ra^{0.277}$	$1.11R^{-0.204}$	$0.0386\widetilde{Ra}^{0.256}$	$0.803R^{-0.0554}$
Simulations (2D DNS)							
Jahn and Reineke [39, 53]	7	$1\cdot10^5$–$1\cdot10^9$	$2.20R^{-0.192}$	$0.0678Ra^{0.238}$			$1.19R^{-0.0854}$
Peckover and Hutchinson [57] (free-slip, steady)	8	$5.1\cdot10^4$–$1.4\cdot10^6$	$0.575R^{-0.104}$	$0.182Ra^{0.116}$			$0.953R^{-0.0752}$
Straus [74] (free-slip, steady)	∞	10^5–$3\cdot10^5$	$1.82R^{-0.217}$	$0.0795Ra^{0.277}$			$1.23R^{-0.100}$
Emara and Kulacki [22]	6.5	$5\cdot10^4$–$5\cdot10^7$	$1.96R^{-0.186}$	$0.0795Ra^{0.229}$			$1.02R^{-0.0672}$
Goluskin and Spiegel [27]	1	10^8–$2\cdot10^{10}$			$1.13R^{-0.200}$	$0.0384\widetilde{Ra}^{0.250}$	

Internal heating was achieved by heating elements in one laboratory experiment [50] and by electric current in the others. Fits to $\overline{T}_{\max}$ and $\mathscr{F}_B$ are computed directly from data for a few studies [27, 28, 47, 84], while for other studies we have computed them from reported fits to N_T and N_B (see text). Simulations employ no-slip boundary conditions, except when specified otherwise

range from $\gamma = 0.220$ to $\gamma = 0.289$. The bottom end of this range is smaller than any exponents found for the ordinary RB1 Rayleigh number, except at very small Pr [29].

3.2.2.2 Mean Temperatures

The quantity $\delta\langle T\rangle$ has been reported only in two numerical studies, and each gives a fit of the form $\delta\langle T\rangle \sim bR^{-\beta}$ for $Pr = 1$. Despite one study being 3D and the other 2D, the growth rates of $\delta\langle T\rangle$ with R are very similar, having exponents of $\beta = 0.204$ in 3D [28] and $\beta = 0.200$ in 2D [27]. This is reminiscent of the Nusselt number in RB convection, which is not much affected by dimensionality unless Pr is small [64, 82].

The dimensional mean temperature, $\delta\langle T\rangle\Delta$, grows with the volumetric heating proportionally to $H^{0.796}$ in 3D and to $H^{0.800}$ in 2D. When the $\delta\langle T\rangle$ fits are restated in the form $\tilde{N} \sim d\widetilde{Ra}^{\delta}$, the exponents are $\delta = 0.256$ in 3D and $\delta = 0.250$ in 2D. These δ values are within the range of Nusselt number growth rates seen in RB1 convection, though they are at the lower end of that range (cf. Sect. 3.1.2). We cannot yet draw comparison with IH3 convection, for which no data on $\delta\langle T\rangle$ have been reported.

3.2.2.3 Asymmetry

The asymmetry between upward and downward heat fluxes in IH1 convection, as quantified by the fraction of heat that flows downward, $\mathscr{F}_B$, seems to have no analogues in our other five IH or RB configurations. First, this fraction changes with R much more slowly than any other integral quantity we have discussed. Second, the R-dependence of $\mathscr{F}_B$ can differ greatly between 2D and 3D, even when Pr is not small. This is because shear, rather than buoyancy, is the mechanism responsible for mixing the cooler lower boundary layer with the warmer interior. The asymmetry is generally greater in 3D than in 2D because the shear-driven mixing, which helps heat escape across the bottom boundary, is less effective in 3D [28].

A particularly simply question without an obvious answer is: as $R \to \infty$, what is the limit of $\mathscr{F}_B$? The extreme possibilities of either 0 or 1/2 seem most likely, although intermediate values are also plausible. In the highest-R simulation data available in 3D, $\mathscr{F}_B$ falls monotonically as R is raised [28]. The highest-R data available in 2D are quite different, except perhaps at large Pr [27, 28]. For instance, in the 2D simulations of Goluskin and Spiegel [27] with $Pr = 1$, the fraction $\mathscr{F}_B$ reaches a minimum of 0.33 near $R = 10^9$ and then increases as R is raised further. This non-monotonic R-dependence in 2D is yet another way that $\mathscr{F}_B$ stands apart from other quantities we have considered.

When $\mathscr{F}_B$ decreases monotonically as R is raised, as in all past 3D studies and some 2D ones, we can seek fits of the form $\mathscr{F}_B \sim eR^{-\varepsilon}$. Table 3.2 summarizes these fits, all of whose decay rates are quite small. The decay rates range from $\varepsilon = 0.0494$

to $\varepsilon = 0.0988$. It remains a mystery whether such decay will continue or reverse at larger R.

The dependence of $\mathscr{F}_B$ on Pr has been examined in two studies [27, 28]. The value of $\mathscr{F}_B$ seems to fall monotonically as Pr is raised, meaning that the asymmetry increases, until saturating at large Pr. The effect of Pr on the asymmetry is fairly strong—stronger than its effect on $\delta\langle T\rangle$ or $\overline{T}_{\max}$.

3.2.3   Scaling Arguments

Several scaling arguments have been put forth to explain the parameter-dependence of mean temperatures in IH convection for both the IH1 and IH3 configurations [15–17, 27]. In RB convection, the Nusselt number displays a wide diversity of scaling behavior in different regions of parameter space [72]. It is likely that the same is true of mean temperatures in IH convection since (inverses of) these temperatures have many parallels to the RB Nusselt number. This remains to be confirmed by a wider exploration of parameter space. If such a diversity of scaling can indeed be found in IH convection, then any broadly applicable scaling arguments must reflect this. For the standard RB1 configuration, the only arguments that attempt to capture the full range of scaling behavior are those put forth by Grossman and Lohse in [29] and subsequent papers (see [72]). The arguments of [29] carry through analogously for IH convection [27]. When the predicted scalings are phrased in terms of N and Ra, or $\tilde{N}$ and $\widetilde{Ra}$, they are the same as the scalings predicted for the Nusselt number in the RB1 case. However, further work on scaling arguments is perhaps premature until data are available across a wider swath of parameter space.

3.3   Future Directions

In the future study of IH convection, the main task accessible to mathematical analysis is proving parameter-dependent bounds on key integral quantities. The only results of this kind are the R-dependent lower bounds on volume-averaged temperatures described in Sect. 2.3. We have conjectured in Sect. 1.6.3 that there should also exist R-dependent upper bounds on the mean convective flux, $\langle wT\rangle$. These would amount to lower bounds on the fraction of heat flowing downward in IH1 and on the mean temperature difference between the boundaries in IH2 and IH3. Bounds are lacking also for the maximum horizontally averaged temperature, $\overline{T}_{\max}$, that has often been measured in IH1 experiments. Bounds depending analytically on the Prandtl number are highly desirable as well.

There is much fertile ground for physical and computational experiments on IH convection. This is especially true for computation since most prior results are several decades old, so modern computers would be able to probe unexplored parameter regimes with relative ease. Neither the IH2 nor IH3 configuration has been simulated in 3D, and the DNS carried out in 2D has not approached the large R that are now computationally accessible. The IH1 configuration has been the subject of two DNS studies in 3D [28, 84], but a much wider exploration of parameter space is called for. The asymmetry between upward and downward heat fluxes in IH1 is particularly hard to predict; even its value as R approaches infinity is not certain. In each of the three IH configurations, simulating a wide range of R and Pr would produce a more global picture of how key integral quantities depend on the control parameters. The complicated parameter-dependence of Nusselt numbers in RB convection [1, 72] suggests that fitting integral quantities to pure powers of R will not suffice.

A combination of mathematical analysis, simulation, and physical experimentation will lead to a better understanding of the three internally heated configurations we have studied in this SpringerBrief. We hope that this, in turn, will lead to a better understanding of more complicated occurrences of IH convection. The many past studies of RB convection should prove useful in guiding future studies of IH convection, and to this end we have described a number of analogies between the two classes of flows. Still, the analogies are not perfect, and some consequences of internal heating cannot be foreseen. Judging by the complexity of RB convection, we expect that these novel aspects of IH convection will remain rich areas of inquiry for many more years.

References

1. Ahlers, G., Grossmann, S., Lohse, D.: Heat transfer and large scale dynamics in turbulent Rayleigh-Bénard convection. Rev. Mod. Phys. **81**(2), 503–537 (2009)
2. Ames, K.A., Straughan, B.: Penetrative convection in fluid layers with internal heat sources. Acta Mech. **85**, 137–148 (1990)
3. Arcidiacono, S., Ciofalo, M.: Low-Prandtl number natural convection in volumetrically heated rectangular enclosures III. Shallow cavity, AR=0.25. Int. J. Heat Mass Transf. **44**, 3053–3065 (2001)
4. Arcidiacono, S., Di Piazza, I., Ciofalo, M.: Low-Prandtl number natural convection in volumetrically heated rectangular enclosures II. Square cavity, AR=1. Int. J. Heat Mass Transf. **44**, 537–550 (2001)
5. Asfia, F.J., Dhir, V.K.: An experimental study of natural convection in a volumetrically heated spherical pool bounded on top with a rigid wall. Nucl. Eng. Des. **163**(3), 333–348 (1996)
6. Bergholz, R.F.: Natural convection of a heat generating fluid in a closed cavity. J. Heat Transfer **102**, 242–247 (1980)
7. Berlengiero, M., Emanuel, K.A., von Hardenberg, J., Provenzale, A., Spiegel, E.A.: Internally cooled convection: A fillip for Philip. Commun. Nonlinear Sci. Numer. Simul. **17**(5), 1998–2007 (2012)
8. Busse, F.H.: Patterns of convection in spherical shells. J. Fluid Mech. **72**(1), 67–85 (1975)
9. Busse, F.H., Riahi, N.: Patterns of convection in spherical shells. Part 2. J. Fluid Mech. **123**, 282–301 (1975)
10. Cartland Glover, G.M., Generalis, S.C.: Pattern competition in homogeneously heated fluid layers. Eng. Appl. Comput. Fluid Mech. **3**(2), 164–174 (2009)
11. Cartland Glover, G., Fujimura, K., Generalis, S.: Pattern formation in volumetrically heated fluids. Chaotic Model. Simul. **1**, 19–30 (2013)

12. Chapman, C.J., Childress, S., Proctor, M.R.E.: Long wavelength thermal convection between non-conducting boundaries. Earth Planet. Sci. Lett. **51**, 362–369 (1980)
13. Chavanne, X., Chilla, F., Castaing, B., Hebral, B., Chabaud, B., Chaussy, J.: Observation of the ultimate regime in Rayleigh-Bénard convection. Phys. Rev. Lett. **79**(19), 3648–3651 (1997)
14. Chen, S., Krafczyk, M.: Entropy generation in turbulent natural convection due to internal heat generation. Int. J. Therm. Sci. **48**(10), 1978–1987 (2009)
15. Cheung, F.B.: Natural convection in a volumetrically heated fluid layer at high Rayleigh numbers. Int. J. Heat Mass Transf. **20**(5), 499–506 (1977)
16. Cheung, F.B.: Heat source-driven thermal convection at arbitrary Prandtl number. J. Fluid Mech. **97**(4), 743–758 (1980)
17. Cheung, F.B., Chawla, T.C.: Complex heat transfer processes in heat-generating horizontal fluid layers. In: Annual review of numerical fluid mechanics and heat transfer, vol. 1, pp. 403–448. Hemisphere, New York (1987)
18. Chillà, F., Schumacher, J.: New perspectives in turbulent Rayleigh-Bénard convection. Eur. Phys. J. E **35**(7), 1–25 (2012)
19. Clever, R.M.: Heat transfer and stability properties of convection rolls in an internally heated fluid layer. J. Appl. Math. Phys. **28**, 585–597 (1977)
20. De la Cruz Reyna, S.: Asymmetric convection in the upper mantle. Geofis. Int. **10**, 49–56 (1970)
21. Di Piazza, I., Ciofalo, M.: Low-Prandtl number natural convection in volumetrically heated rectangular enclosures I. Slender cavity, AR=4. Int. J. Heat Mass Transf. **43**, 3027–3051 (2000)
22. Emara, A.A., Kulacki, F.A.: A numerical investigation of thermal convection in a heat-generating fluid layer. J. Heat Transf. **102**, 531–537 (1980)
23. Farouk, B.: Turbulent thermal convection in an enclosure with internal heat generation. J. Heat Transf. **110**(1), 126–132 (1988)
24. Fiedler, H.E., Wille, R.: Turbulente freie konvektion in einer horizontalen flüssigkeitsschicht mit volumen-wärmequelle. In: Proceeding of 4th International Heat Transfer Conference (1970)
25. Filippov, A.S.: Numerical simulation of experiments on turbulent natural convection of heat generating liquid in cylindrical pool. J. Eng. Thermophys. **20**(1), 64–76 (2011)
26. Getling, A.V.: Rayleigh-Bénard convection: structures and dynamics. World Scientific Publishing Co (1998)
27. Goluskin, D., Spiegel, E.A.: Convection driven by internal heating. Phys. Lett. A **377**(1-2), 83–92 (2012)
28. Goluskin, D., van der Poel, E.P.: Penetrative internally heated convection in two and three dimensions. Submitted. (2015)
29. Grossmann, S., Lohse, D.: Scaling in thermal convection: a unifying theory. J. Fluid Mech. **407**, 27–56 (2000)
30. Grötzbach, G.: Turbulent heat transfer in an internally heated fluid layer. In: The Third International Symposium on Refined Flow Modelling and Turbulence Measurements, vol. 2, p. 8. Tokyo (1988)
31. Hartlep, T., Busse, F.H.: Convection in an internally cooled fluid layer heated from below. Technical Representation, Center for Turnulence Research (2006)
32. He, X., Funfschilling, D., Nobach, H., Bodenschatz, E., Ahlers, G.: Transition to the ultimate state of turbulent Rayleigh-Bénard convection. Phys. Rev. Lett. **108**, 024502 (2012)
33. Hewitt, J.M., McKenzie, D.P., Weiss, N.O.: Large aspect ratio cells in two-dimensional thermal convection. Earth Planet. Sci. Lett. **51**, 370–380 (1980)
34. Horvat, A., Kljenak, I., Marn, J.: Two-dimensional large-eddy simulation of turbulent natural convection due to internal heat generation. Int. J. Heat Mass Transf. **44**(21), 3985–3995 (2001)
35. Houseman, G.: The dependence of convection planform on mode of heating. Nature **332**, 346–349 (1988)
36. Ichikawa, H., Kurita, K., Yamagishi, Y., Yanagisawa, T.: Cell pattern of thermal convection induced by internal heating. Phys. Fluids **18**(3), 038101 (2006)

37. Ingersoll, A.P., Porco, C.C.: Solar heating and internal heat flow on Jupiter. Icarus **35**, 27–43 (1978)
38. Ishiwatari, M., Takehiro, S.I., Hayashi, Y.Y.: The effects of thermal conditions on the cell sizes of two-dimensional convection. J. Fluid Mech. **281**, 33–50 (1994)
39. Jahn, M., Reineke, H.H.: Free convection heat transfer with internal heat sources, calculations and measurements. In: Proceedings of 5th International Heat Transfer Conference, pp. 74–78. Tokyo (1974)
40. Jaupart, C., Brandeis, G.: The stagnant bottom layer of convecting magma chambers. Earth Planet. Sci. Lett. **80**, 183–199 (1986)
41. Jaupart, C., Brandeis, G., Allègre, C.J.: Stagnant layers at the bottom of convecting magma chambers. Nature **308**, 535–538 (1984)
42. Joseph, D.D.: Subcritical Instability and Exchange of Stability in a Horizontal Fluid Layer. Phys. Fluids **11**(1968), 903–904 (1968)
43. Joseph, D.D., Shir, C.C.: Subcritical convective instability: Part 1. Fluid layers. J. Fluid Mech. **26**(4), 753–768 (1966)
44. Kolmychkov, V.V., Mazhorova, O.S., Shcheritsa, O.V.: Numerical study of convection near the stability threshold in a square box with internal heat generation. Phys. Lett. A **377**, 2111–2117 (2013)
45. Kondratenko, P.S., Nikolski, D.V., Strizhov, V.F.: Free-convective heat transfer in fluids with non-uniform volumetric heat generation. Int. J. Heat Mass Transf. **51**(7-8), 1590–1595 (2008)
46. Kulacki, F.A., Emara, A.A.: Steady and transient thermal convection in a fluid layer with uniform volumetric energy sources. J. Fluid Mech. **83**(2), 375–395 (1977)
47. Kulacki, F.A., Goldstein, R.J.: Thermal convection in a horizontal fluid layer with uniform volumetric energy sources. J. Fluid Mech. **55**(02), 271–287 (1972)
48. Kulacki, F.A., Nagle, M.E.: Natural convection in a horizontal fluid layer with volumetric energy sources. J. Heat Transf. **97**, 204–211 (1975)
49. Kulacki, F.A., Richards, D.E.: Natural convection in plane layers and cavities with volumetric energy sources. In: Natural Convection: Fundamentals and Applications, pp. 179–254. Hemisphere, New York (1985)
50. Lee, S.D., Lee, J.K., Suh, K.Y.: Boundary condition dependent natural convection in a rectangular pool with internal heat sources. J. Heat Transf. **129**(5), 679–682 (2007)
51. Liu, H., Zou, C., Shi, B., Tian, Z., Zhang, L., Zheng, C.: Thermal lattice-BGK model based on large-eddy simulation of turbulent natural convection due to internal heat generation. Int. J. Heat Mass Transf. **49**, 4672–4680 (2006)
52. Lohse, D., Xia, K.Q.: Small-scale properties of turbulent Rayleigh-Bénard convection. Annu. Rev. Fluid Mech. **42**(1), 335–364 (2010)
53. Mayinger, F., Jahn, M., Reineke, H.H., Steinberner, U.: Examination of thermohydraulic processes and heat transfer in a core melt. Technical Representation, Hannover Technical University, Hannover, Germany (1975)
54. McKenzie, D.P., Roberts, J.M., Weiss, N.O.: Convection in the earth's mantle: towards a numerical simulation. J. Fluid Mech. **62**(3), 465–538 (1974)
55. Nourgaliev, R.R., Dinh, T.N., Sehgal, B.R.: Effect of fluid Prandtl number on heat transfer characteristics in internally heated liquid pools with Rayleigh numbers up to 10^{12}. Nucl. Eng. Des. **169**, 165–184 (1997)
56. Olwi, I.A., Kulacki, F.A.: Numerical simulation of the transient convection process in a volumetrically heated fluid layer. In: Proceeding of ASME, p. 185 (1995)
57. Peckover, R.S., Hutchinson, I.H.: Convective rolls driven by internal heat sources. Phys. Fluids **17**(7), 1369–1371 (1974)
58. Ralph, J.C., Roberts, D.N.: Free convection heat transfer measurements in horizontal liquid layers with internal heat generation. Technical Representation, UKAEA (1974)
59. Ralph, J.C., McGreevy, R., Peckover, R.S.: Experiments in tubulent thermal convection driven by internal heat sources. In: Spalding, D.B., Afgan, N. (eds.) Heat Transfer and Turbulent Buoyant Convection: Studies and Applications for Natural Environment, Buildings, Engineering Systems, pp. 587–599. Hemisphere, New York (1977)

60. Riahi, N.: Nonlinear convection in a horizontal layer with an internal heat source. J. Phys. Soc. Japan **53**(12), 4169–4178 (1984)
61. Riahi, D.N., Busse, F.H.: Pattern generation by convection in spherical-shells. J. Appl. Math. Phys. **39**, 699–712 (1988)
62. Roberts, P.H.: Convection in a self-gravitating fluid sphere. Mathematika **12**, 128 (1965)
63. Roberts, P.H.: On the thermal instability of a rotating-fluid sphere containing heat sources. Philos. Trans. R. Soc. A **263**, 93–117 (1968)
64. Schmalzl, J., Breuer, M., Hansen, U.: On the validity of two-dimensional numerical approaches to time-dependent thermal convection. Europhys. Lett. **67**(3), 390–396 (2004)
65. Schubert, G., Glatzmaier, G.A., Travis, B.: Steady, three-dimensional, internally heated convection. Phys. Fluids A **5**(8), 1928–1932 (1993)
66. Schwiderski, E.W., Schwab, H.J.A.: Convection experiments with electrolytically heated fluid layers. J. Fluid Mech. **48**(4), 703–719 (1971)
67. Shi, B.C., Guo, Z.L.: Thermal lattice BGK simulation of turbulent natural convection due to internal heat generation. Int. J. Mod. Phys. B **17**(2), 173–177 (2003)
68. Siggia, E.D.: High Rayleigh number convection. Annu. Rev. Fluid Mech. **26**, 137–168 (1994)
69. Sotin, C., Labrosse, S.: Three-dimensional thermal convection in an iso-viscous, infinite Prandtl number fluid heated from within and from below: applications to the transfer of heat through planetary mantles. Phys. Earth Planet. Inter. **112**, 171–190 (1999)
70. Sparrow, E.M., Goldstein, R.J., Jonsson, V.K.: Thermal instability in a horizontal fluid layer: effect of boundary conditions and non-linear temperature profile. J. Fluid Mech. **18**(04), 513–528 (1964)
71. Steinberner, U., Reineke, H.H.: Turbulent buoyancy convection heat transfer with internal heat sources. In: Proceeding of 6th International Heat Transfer Conference, vol. 2, pp. 305–310 (1978)
72. Stevens, R.J.A.M., van der Poel, E.P., Grossmann, S., Lohse, D.: The unifying theory of scaling in thermal convection: the updated prefactors. J. Fluid Mech. **730**, 295–308 (2013)
73. Straughan, B.: Continuous dependence on the heat source and non-linear stability for convection with internal heat generation. Math. Methods Appl. Sci. **13**, 373–383 (1990)
74. Straus, J.M.: Penetrative convection in a layer of fluid heated from within. Astrophys. J. **209**, 179–189 (1976)
75. Takahashi, J., Tasaka, Y., Murai, Y., Takeda, Y., Yanagisawa, T.: Experimental study of cell pattern formation induced by internal heat sources in a horizontal fluid layer. Int. J. Heat Mass Transf. **53**(7-8), 1483–1490 (2010)
76. Tasaka, Y., Takeda, Y.: Effects of heat source distribution on natural convection induced by internal heating. Int. J. Heat Mass Transf. **48**(6), 1164–1174 (2005)
77. Tasaka, Y., Kudoh, Y., Takeda, Y., Yanagisawa, T.: Experimental investigation of natural convection induced by internal heat generation. J. Phys. Conf. Ser. **14**, 168–179 (2005)
78. Thirlby, R.: Convection in an internally heated layer. J. Fluid Mech. **44**(04), 673–693 (1970)
79. Tritton, D.J., Zarraga, M.N.: Convection in horizontal layers with internal heat generation. Experiments. J. Fluid Mech. **30**(01), 21–31 (1967)
80. Tveitereid, M.: Thermal convection in a horizontal fluid layer with internal heat sources. Int. J. Heat Mass Transf. **21**, 335–339 (1978)
81. Tveitereid, M., Palm, E.: Convection due to internal heat sources. J. Fluid Mech. **76**(03), 481–499 (1976)
82. van der Poel, E.P., Stevens, R.J.A.M., Lohse, D.: Comparison between two- and three-dimensional Rayleigh-Bénard convection. J. Fluid Mech. **736**, 177–194 (2013)
83. Vel'tishchev, N.F.: Convection in a horizontal fluid layer with a uniform internal heat source. Fluid Dyn. **39**(2), 189–197 (2004)
84. Wörner, M., Schmidt, M., Grötzbach, G.: Direct numerical simulation of turbulence in an internally heated convective fluid layer and implications for statistical modeling. **35**(6), 773–797 (1997)